SOCIETY FOR THE STUDY OF HUMAN BIOLOGY SYMPOSIUM SERIES: 31

Fertility and resources

PUBLISHED SYMPOSIA OF THE SOCIETY FOR THE STUDY OF HUMAN BIOLOGY

1 The Scope of Physical Anthropology and its Place in Academic Studies ED. D. F. ROBERTS AND J. S. WEINER
2 Natural Selection in Human Populations ED. D. F. ROBERTS AND G. A. HARRISON
3 Human Growth ED. J. M. TANNER
4 Genetical Variation in Human Populations ED. G. A. HARRISON
5 Dental Anthropology ED. D. R. BROTHWELL
6 Teaching and Research in Human Biology ED. G. A. HARRISON
7 Human Body Composition, Approaches and Applications ED. D. R. BROTHWELL
8 Skeletal Biology of Earlier Human Populations ED. J. BROZEK
9 Human Ecology in the Tropics ED. J. P. GARLICK AND R. W. J. KEAY
10 Biological Aspects of Demography ED. W. BRASS
11 Human Evolution ED. M. H. DAY
12 Genetic Variation in Britain ED. D. F. ROBERTS AND E. SUNDERLAND
13 Human Variation and Natural Selection ED. E. F. ROBERTS (Penrose Memorial Volume reprint)
14 Chromosome Variation in Human Evolution ED. A. J. BOYCE
15 Biology of Human Foetal Growth ED. D. F. ROBERTS
16 Human Ecology in the Tropics ED. J. P. GARLICK AND R. W. J. DAY
17 Physiological Variation and its Genetic Base ED. J. S. WEINER
18 Human Behaviour and Adaptation ED. N. J. BLURTON JONES AND V. REYNOLDS
19 Demographic Patterns in Developed Societies ED. R. W. HIORNS
20 Disease and Urbanisation ED. E. J. CLEGG AND J. P. GARLICK
21 Aspects of Human Evolution ED. C. B. STRINGER
22 Energy and Effort ED. G. A. HARRISON
23 Migration and Mobility ED. A. J. BOYCE
24 Sexual Dimorphism ED. F. NEWCOMBE *et al.*
25 The Biology of Human Ageing ED. A. H. BITTLES AND K. J. COLLINS
26 Capacity for Work in the Tropics ED. K. J. COLLINS AND D. F. ROBERTS
27 Genetic Variation and its Maintenance ED. E. F. ROBERTS AND G. F. DE STEFANO
28 Human Mating Patterns ED. C. G. N. MASCIE-TAYLOR AND A. J. BOYCE
29 The Physiology of Human Growth ED. J. M. TANNER AND M. A. PREECE
30 Diet and Disease ED. G. A. HARRISON AND J. WATERLOW
31 Fertility and Resources ED. J. LANDERS AND V. REYNOLDS

Numbers 1–9 were published by Pergamon Press, Headington Hill Hall, Headington, Oxford OX3 0BY. Numbers 10–24 were published by Taylor & Francis Ltd, 10–14 Macklin Street, London WC2B 5NF. Further details and prices of back-list numbers are available from the Secretary of the Society for the Study of Human Biology.

Fertility and Resources

31st Symposium Volume of the Society for the Study of Human Biology

EDITED BY

JOHN LANDERS
Department of Anthropology
University College London

AND

VERNON REYNOLDS
Department of Biological Anthropology
University of Oxford

CAMBRIDGE UNIVERSITY PRESS
Cambridge
New York Port Chester Melbourne Sydney

CAMBRIDGE UNIVERSITY PRESS
Cambridge, New York, Melbourne, Madrid, Cape Town, Singapore, São Paulo, Delhi

Cambridge University Press
The Edinburgh Building, Cambridge CB2 8RU, UK

Published in the United States of America by Cambridge University Press, New York

www.cambridge.org
Information on this title: www.cambridge.org/9780521103039

First published 1990
This digitally printed version 2009

A catalogue record for this publication is available from the British Library

Library of Congress Cataloguing in Publication data

Society of the Study of Human Biology. Symposium (31st: 1989:
Oxford University)
Fertility and resources: 31st Symposium volume of the Society for
the Study of Human Biology / edited by J. Landers and V. Reynolds.
p. cm. – (Society for the Study of Human Biology symposium
series; 31)
Symposium held at Oxford University, April 10–11, 1989.
Includes index.
ISBN 0 521 39526 7
1. Fertility, Human – Congresses. 2. Fertility, Human – Cross-
cultural studies – Congresses. 3. Man – Influence of environment –
Congresses. I. Landers, J. (John), 1952– . II. Reynolds,
Vernon. III. Title. IV. Series.
[DNLM: 1. Anthropology, Cultural – congresses. 2. Fertility –
congresses. W1 SO861 v. 31/WP 565 S678f 1989]
GN241.S63 1990
304.6'32—dc20 90-1833 CIP

ISBN 978-0-521-39526-7 hardback
ISBN 978-0-521-10303-9 paperback

Contents

List of contributors

Dr A. H. Bittles
King's College, Strand, London. WC2R 2LS.

Dr E. Boserup
Casa Campagnola, Nevadore, 6614 Brissago, Switzerland.

Dr J. Cleland
Centre for Population Studies, London School of Hygiene and Tropical Medicine, 99 Gower Street, London, WC1E 6AZ.

A. Radha Rama Devi
Health Centre, Indian Institute of Science, Bangalore, India

Dr R. I. M. Dunbar
Department of Anthropology, University College London, Gower Street, London. WC1E 6BT.

Dr P. M. Everson
Department of Anthropology, University of Washington, Seattle, WA 98195, USA

Dr A. G. Hill
The Centre for Population Studies, London School of Hygiene and Tropical Medicine. 99 Gower Street, London, WC1E 6AZ

Dr P. R. A. Hinde
Department of Social Statistics, University of Southampton, Southampton, SO9 5NH.

Dr J. Landers
Department of Anthropology, University College London, Gower Street, London. WC1E 6BT.

Dr J. Lourie
Milton Keynes Hospital, Milton Keynes, Beds.

Dr V. Mea
Medical Faculty, University of Papua New Guinea, Box 320, University Post Office, Port Moresby, Papua New Guinea.

1 *Introduction to Fertility and Resources*

JOHN LANDERS AND VERNON REYNOLDS

The relationship between the fertility of organisms and their food supply and other resources has been very extensively studied and forms the subject matter of the science of animal ecology. Studies of this relationship in human beings have had to meet the complication that the food supply of humans is not a fixed resource, but is subject to human manipulation in a number of ways. In addition, food resources, like resources generally, are subject to cultural definition and people in one society may regard as inedible foods that are highly prized elsewhere. While biologists have sought for, and often found, clear relationships between food quality and abundance on the one hand and patterns of fertility on the other, anthropologists, sociologists and demographers have shown that human manipulation of both food production and fertility levels make theory formation much harder in the human case.

In this book a number of biologists, anthropologists and demographers examine how patterns of reproduction are related to the availability of food and other scarce resources in a variety of conditions and circumstances. Dunbar begins by looking at the situation in non-human primates. He points to the importance of social factors as well as the more commonly considered ecological ones. Group size and composition, sex ratio, dominance rank, and membership of coalitions all affect the fertility of female primates. A particular set of proximate mechanisms at the physiological level, especially the actions of prolactin and brain opiates, appear to underlie primate reproductive variance.

Rosetta's analysis of seasonality in a West African subsistence economy emphasizes the phenomenon of food shortage at just the time of year when the harvest occurs. Like Dunbar, Rosetta argues that the proximate mechanisms leading to reproductive suppression at the time of food scarcity and hard work are the outcome of changes in CNS and peripheral functioning.

Taufa turns our attention to a situation of dramatic change in ecological circumstances. The Wopkaimin people of Papua New Guinea have been subject to a complete alteration in their lifestyle as a result of the construction of the Ok Tedi mine in their area of the highlands. Taufa shows how this has led to a great increase in fertility and to a fall in the duration of birth intervals, despite an increase in the practice of contraception. The underlying factors appear to be improvements in health care together with the introduction of a cash economy.

A more gradual change in the other direction is documented in Stevenson and Everson's chapter on the Mennonites of Kansas and Nebraska. Immigrants from the Ukraine, their family size has steadily decreased in the last 100 years from an initially high level towards that of the host community. The authors find Caldwell's 'wealth flows' model, which explains fertility in terms of a reversal in the direction of such flows between generations, useful in accounting for the changes they document.

Bittles, Devi and Rao, in a study of fertility in Karnataka, show how cultural factors interact with biological ones. In this part of India, Hindus and Moslims practise high levels of consanguineous marriages. Moslim fertility is the highest. Perhaps attributable to the longer reproductive span of Moslims, their poor socio-economic status and education, and attitudes to contraception and abortion.

Hinde applies household economics theory to the decline of fertility in later nineteenth century rural England. Nationally the north and west contrasted with southern and eastern England, the former having relatively low nuptiality combined with high marital fertility, whilst the latter had the reverse configuration. Hinde links this contrast to the varying incidence of residential farm service, complementing his national survey with a detailed analysis of smaller areas in Norfolk and Shropshire. Living standards apparently rose in both areas, until the early 1870s when the effects of declining agricultural wages were compounded by the imposition of compulsory education and falling demand for female and child labour. Pre-conditions for fertility decline developed gradually after 1850, but its actual occurrence required the 'trigger' of sudden (if temporary) economic pressure.

The decline of marital fertility among eighteenth century London Quakers is examined by Landers. Like other so-called 'precursor' populations the London Quakers show signs of voluntary fertility control substantially before the wider society. Their behaviour, however, deviated from that predicted by the widely accepted parity-specific model of family limitation since there was a substantial slowing of the tempo of fertility early in marriage, particularly among women marrying in the later twenties. The author outlines a simple model attributing the behaviour of the older brides to a concern for existential, rather than completed, family size. The same criterion, however, results in parity specific 'stopping' behaviour among younger brides. Parity-specific fertility control can thus arise as an unintended consequence of a reaction to current conditions and does not necessarily imply the alteration of fertility behaviour with respect to desired completed family size.

Boserup considers the relationship between population growth, innovation and resource exploitation. Population growth generally favours productivity increase which, through the upgrading of human resources and technology, can counter-balance the decline of physical resources per capita brought on by such growth. However, some technological changes damage the physical environment, human resources, or both. While such effects can often be substantially reduced, producers frequently avoid the necessary costs involved and competitive considerations lead governments to reject the necessary legislation or public investments. Political pressures for internationally coordinated control are accumulating, however, and strict adherence to the principle of national sovereignty is incompatible with a high technology world.

The influence of economic factors, cultural factors and government interventions on fertility declines in developing countries is examined by Cleland. Theories of fertility decline have generally argued that aspects of structural modernization reduce the need for, and advantage of, high fertility, but the empirical record shows that such factors fail to explain either historical European fertility declines, or more recent reductions in developing countries. Ill-defined and poorly understood cultural factors appear to exert an additional critical influence. Government policies and interventions to reduce fertility complicate interpretations further and their impact remains highly controversial. The author argues that, ironically, the most widely applauded programmes may have achieved the least, whilst others have affected fertility, albeit modestly.

Hill's paper reviews the question of fertility decline in the contemporary Third World. A variety of different sources have shown these to be substantial, although less so in sub-Saharan Africa. Pre-decline levels of fertility were themselves, however, substantially below the theoretical physiological maximum as a result of lengthy periods of breast feeding. The effect of modernization has generally been to reduce periods of lactation and thus the impact of this biological control over fecundity. Hence in countries of currently low contraceptive prevalence fertility trends will reflect the balance between declining breastfeeding and the spread of contraception.

The origins and extent of the distinctive 'north western European marriage pattern', which combined late marriage for women with a substantial proportion remaining permanently unmarried, form the subject of Richard Smith's contribution. In a series of influential publications the social anthropologist Jack Goody has argued that the pattern followed the establishment of a distinctive system of inheritance under the influence of the medieval Church. Smith considers these claims, concluding however that the contrast between northern and southern European marriage patterns, together with the latter's similarity to those of the Islamic Mediterranean, suggest that more plausible explanations might be sought in the existence of culturally determined regional variations in systems of household structure and formation.

2 *Environmental and Social Determinants of Fecundity in Primates*

R. I. M. DUNBAR

During the past decade, detailed data have become available on the lifehistory variables of free-living primates. Although the number of species for which such data are available is still relatively small (and nearly all are Old World monkeys), nonetheless we now know enough to begin to piece together a more coherent picture of the factors that influence fecundity in primates.

Aside from such intrinsic factors as age, two extrinsic factors emerge as important determinants of female fecundity: these are the environmental conditions (principally the availability of food) and the social and demographic circumstances in which an animal finds itself. In this paper, I review the ways in which these factors influence two aspects of female fecundity, namely age at first reproduction and the birth rate itself. I then consider the proximate mechanisms that mediate these effects.

It should be noted that, throughout this paper, I follow biological usage in using the term *fecundity* to refer to the birth rate. Human demographers conventionally use the term *fertility* in this context, reserving the term fecundity for the ability to conceive (Cox, 1976). Since reproductive physiologists refer to the ability to conceive as fertility, population biologists have tended to follow suit in order to maintain consistency of usage, and the term fecundity has therefore come to refer to the subsequent occurrence of births (see Ricklefs, 1973; Caughley, 1977; but for an exception, see Wilson & Bossert, 1971).

Determinants of female fecundity

Intrinsic factors

Most studies of wild primates have documented variations in fecundity with age. Following a period of post-menarche juvenile sterility, birth rates increase to reach a peak some two to four years after puberty. While a number of studies have documented a decline in fecundity as females continue to age (baboons: Dunbar, 1980; Strum & Western, 1982), some studies have found that birth rates remain more or less constant once reaching their peak in mid-prime (e.g. macaques: Silk, 1988; vervets: Cheney *et al.*, 1988).

A second intrinsic factor influencing birth rates in primates is infant survival. In non-seasonally breeding species, the death of a suckling infant leads to the mother returning to reproductive condition within a few weeks. Reduced inter-birth intervals following the postnatal death of an infant have been documented in colobines (Winkler *et al.*, 1984), macaques (Hadidian & Bernstein, 1979), baboons (Altmann *et al.*, 1977; Dunbar, 1980a) and apes (Goodall, 1983; Stewart 1988).

Environmental factors

Comparisons of age at first reproduction in wild and captive populations of the same species invariably reveal that animals reproduce earlier in captivity where food is normally more abundant (e.g. various Old World monkeys: Gautier-Hion & Gautier, 1976; Hadidian & Bernstein, 1979; Rowell & Richards, 1979; chimpanzee: Goodall, 1983).

In wild populations that are provisioned, females tend to reproduce for the first time at about the same age as they do in captivity (e.g. macaques: Mori, 1979; Fa, 1986). If provisioning is subsequently withdrawn, the age at first reproduction is delayed (macaques: Mori, 1979; Sugiyama & Ohsawa, 1982). Mori (1979) suggested that this delay in the timing of puberty correlated with a slower growth rate: females seemed to undergo puberty at about the same body weight under provisioned and non-provisioned conditions, but to take longer to achieve that body weight when food was less readily available.

Within populations, age at first reproduction has been shown to vary in relation to habitat productivity in a number of studies. Strum & Western (1982), for example, found that the mean age at first reproduction in one group of baboons (which varied between 5.8 and 7.7 years in different years) was significantly correlated with the herbivore biomass: the greater the competition for the limited food resources within the group's home range, the later was the mean age at which maturing females reproduced for the first time. Cheney *et al.* (1988) found that the mean age at first birth was 6–12 months earlier in two groups of vervet monkeys occupying ecologically richer territories than it was in a group occupying the poorest territory.

Inter-birth intervals are clearly the main factor determining a female's fecundity once past puberty. Of the three components that make up the inter-birth interval, the most variable is the length of the period of postpartum amenorrhoea. Many primates are seasonal breeders, but among those that are not, food availability seems to be an important determinant of the length of the period of amenorrhoea.

In captivity, where food supplies are both constant and relatively plentiful, inter-birth intervals of 12–18 months are common in Old World monkeys (Hadidian & Bernstein, 1979; Rowell & Richards, 1979; Smith, 1982; Fairbanks & McGuire, 1984). Comparably high rates of reproduction are found in provisioned populations of free-living macaques (e.g. Drickamer, 1974; Mori, 1979; Fa, 1986). Natural populations are rarely able to achieve such high fecundity: inter-birth intervals of 18–24 months are more typical for most free-ranging Old World monkey populations (Altmann *et al.*, 1977; Takahata, 1980; Strum & Western, 1982; Ohsawa & Dunbar, 1984; Winkler *et al.*, 1984; Bulger & Hamilton, 1987).

The influence of foraging conditions in determining the length of the period of postpartum amenorrhoea can be demonstrated in several ways. The mean inter-birth interval was 17.9 months in the Koshima Japanese macaque population during the period when the population was provisioned, for example, but averaged 37.5 months in the six years following the withdrawal of provisioning. Similarly, van Schaik & van Noordwijk (1985) reported a correlation between fruit production, female condition and the birth rate in longtailed macaques in Indonesia.

A number of studies have reported correlations between fecundity and various indices of habitat quality. Dunbar & Sharman (1983) found a relationship between fecundity and rainfall (normally a reliable index of habitat productivity) in 18 populations of baboons occupying a wide range of habitat types throughout sub-Saharan Africa. Similar results have been reported for macaques (Takahata, 1980; Menard *et al.*, 1985). Similarly, differences in fecundity have been reported between individual groups of both vervet and colobus monkeys occupying adjacent territories that differ significantly in primary biomass and productivity (Lee, 1987; Dunbar, 1987).

One factor other than productivity that may be expected to influence the availability of food within a habitat is the effects of competition from other animals. Ecological competitors may be either of the same or different species. Bulger & Hamilton (1987) found that inter-birth intervals were shortened by about six months when the population density of a baboon group was at its lowest. This correlated with a significant increase in female body weight at low densities. The overall density of ecological competitors may also be expected to have a significant impact. Strum & Western (1982), for example, found that birth rates in one group of baboons over a ten year period were inversely correlated with the total herbivore biomass occupying their range area.

One exception to this pattern has been a *negative* relationship between fecundity and both rainfall and altitude reported for gelada living at high altitudes (2000–4000 m in Ethiopia) (Dunbar, 1980a; Ohsawa & Dunbar, 1984). In this case, it could be shown that low temperatures resulted in high rates of foetal loss during unusually wet years and/or at very high altitudes. Under such conditions, it seems that females are unable to carry the energetic load imposed by pregnancy when deteriorating circumstances impose high thermoregulatory costs.

Social factors

Fecundity has been found to correlate with at least four aspects of social structure in primates. These are: group size and composition, the sex ratio, dominance rank and membership of coalitionary alliances.

Increasing group size generally incurs increasing costs for members of the group in terms of longer day journeys, more frequent disruptions of activities and elevated aggression and stress levels (see van Schaik, 1983; Dunbar, 1987). These can all be expected to influence fecundity adversely. Van Schaik (1983) analysed data from 27 populations of New and Old World monkeys and found that the mean number of immatures per female was negatively correlated with group size in 22 cases. Similarly, Dunbar (1987) found that the number of immatures per female was inversely related to the number of adult females in the group in four East African populations of colobus monkeys. More detailed analysis of data from one of these populations revealed that this was mainly due to the fact that large groups tended to contain many adult males and that the presence of several males increased general levels of aggression among the group's members. Rates of male–male fighting, in particular, increased significantly when a female came into oestrus (Dunbar & Dunbar, 1976), suggesting that it is sexual competition among males that is the root cause in this case. Silk (1988) has also noted an inverse relationship between the birth rate and the number of females in a captive group of bonnet macaques.

Silk (1988) also found that the adult sex ratio was correlated with fecundity independently of group size: the more female-biased the sex ratio, the lower the birth rate. Dunbar & Sharman (1983) reported a similar negative relationship between fecundity and sex ratio in 18 populations of free-living baboons.

In both these cases, it is assumed that all females in the group are affected equally. There is, however, considerable evidence to suggest that this is not always necessarily so. When females are organized in a dominance hierarchy, low-ranking females may bear a disproportionate share of the costs. Correlations between dominance rank and birth rate have been reported from a number of wild and semi-free-ranging primate populations (e.g. tamarins: Garber *et al.*, 1984; vervets: Whitten, 1983; Fairbanks & McGuire, 1984; macaques: Drickamer, 1973; Wilson *et al.*, 1978; Takahata, 1980; Gouzoules *et al.*, 1982; Dittus, 1986; baboons: Dunbar, 1980b; 1989b; for a more general review, see Harcourt, 1987). Although the claim that fedundity is higher in high-ranking females has been disputed (see for example Dolhinow *et al.*, 1979; Wolfe, 1984; Cheney *et al.*, 1988), there seems no good reason to doubt that this relationship does hold at least sometimes. In addition, the daughters of high-ranking females have been found to undergo menarche significantly earlier than those of lower ranking famales in at least one population of free-ranging primates (baboons: Altmann *et al.*, 1988).

Undoubtedly, the least equivocal cases are those concerning marmosets and tamarins (the South American Callitrichidae). Studies of both wild and captive populations have generally reported that only one female in each group breeds (Abbott & Hearn, 1978; Goldizen, 1987). Though groups in which two females have both bred have been noted in wild populations (Goldizen, 1987), it seems that, in most cases, subordinate females are completely reproductively suppressed. Indeed, detailed studies of captive animals indicate that maturing daughters will not even undergo menarche if they remain in their natal groups (Abbott & Hearn, 1978). This phenomenon is unusual in so far as it is found only among the callitrichids where it seems to be associated with the evolution of monogamy, twinning and biparental care (i.e. both parents are involved in rearing the young).

The fourth social factor is the influence that coalitionary alliances can have on the reproductive rates of individual females. Primates are above all social animals and the core of their sociality lies in the formation of social alliances that are designed to protect their members against precisely the kinds of problems of poor reproductive performance that are created by low rank and large group size (Harcourt, 1987; Dunbar, 1988; 1989a).

The functional significance of high-ranking allies in terms of their ability both to reduce the harassment received by a female and to give her access to better food sources has been amply attested to in a wide range of species (e.g. vervets: Cheney, 1983; macaques: Datta, 1983; baboons: Smuts, 1985; Dunbar, 1984). The support and/or protection of high-ranking allies can greatly improve a low-ranking female's reproductive rate, as well as her offspring's chances of surviving to breeding age. Among the gelada, for example, females who are members of coalitions have birth rates that are, on average, 80% higher than those of females of the same age who are not members of a coalition (Dunbar, 1989b). As a result, high-ranking matrilines (which can inevitably form more powerful coalitions) often have higher birth rates and therefore grow faster (e.g. macaques: Mori, 1975; Sade *et al.*, 1976; baboons: Dunbar, 1984).

Proximate mechanisms

Four mechanisms have been proposed as explanations for the proximate control of fertility in primates. These are: 1. animals that are able to gain exclusive access to better quality food sources are consequently in better condition for breeding; 2. harassment and/or high levels of aggression induce physiological stress which then suppresses reproductive function; 3. low-ranking females are mated less often than high-ranking ones because higher-ranking individuals interfere in their relationships with the male(s); and 4. a high sucking frequency by the infant suppresses the mother's return to reproductive condition.

Evidence in favour of the food-access hypothesis rests largely on the demonstration that high-ranking females weigh more (Mori, 1979; Whitten, 1983), have a higher intake of food (Whitten, 1983; Janson, 1985), feed more efficiently (Whitten, 1983; Janson, 1985) or are less likely to die during times of food shortage (Dittus, 1979; Cheney *et al.*, 1988). In reviewing the evidence for this hypothesis, Harcourt (1987) concludes that there is indeed a correlation between dominance rank and fecundity, but only when food is clumped: under these conditions, high-ranking females are able to monopolize access to the best food sources, thereby gaining a reproductive advantage. This can be demonstrated by showing that a correlation between rank and fecundity occurs 1. in captivity but not the wild in a given species (e.g. vervets: Fairbanks & McGuire, 1984 versus Cheney *et al.*, 1988), 2. in wild groups only during periods when the animals were being provisioned (e.g. macaques: Mori, 1979; Sugiyama & Ohsawa, 1982) or 3. in only those free-ranging populations where food sources are more clumped (vervets: Whitten, 1983; capuchins: Janson, 1985). This explanation clearly ties in well with Frisch's (1978; 1982) claim that nutritional condition is the critical factor influencing fertility in human females.

The suggestion that socially-induced stress can adversely affect a female's reproductive physiology, thereby reducing her fertility, is based mainly on laboratory studies. There is considerable experimental evidence to show that females at the bottom of a hierarchy receive more aggression, have higher endogenous opiate levels and suffer from a higher frequency of anovulatory menstrual cycles (Bowman *et al.*, 1978; Abbott *et al.*, 1986). In some species (notably callitrichid primates), subordinate females are so completely suppressed that they will not even undergo puberty while living in groups with other females (Abbott, 1984; 1987).

While it is now generally accepted that the reproductive suppression found in monogamous callitrichids is due to stress-related social processes, there remains some debate as to whether the poor reproductive performance of low-ranking females in large groups of polygamously mating baboons and macaques is due to socially-induced stress or to reduced access to high quality food sources (see Harcourt, 1987). So far, only one field study of free-ranging primates has demonstrated unequivocally that stress rather than food is the causal factor (gelada: Dunbar, 1989b). In this case, a direct test between the predictions of the two hypotheses came down decisively in favour of socially-induced stress as the cause of poor reproductive performance.

Other evidence can, however, be adduced in support of the stress hypothesis. Silk (1988) found that the annual birth rate correlated negatively both with the absolute number of adult females in the group and with the adult sex ratio over a 16 year period in a captive group of macaques. Since food, water and health care were available on an *ad libitum* basis, it is difficult to see how this effect can have been mediated through access to food.

Other studies have reported a negative correlation between birth rates and stress levels (macaques: Sackett *et al.*, 1975; Wolfe, 1979; Fa, 1986; vervets: Fairbanks & McGuire, 1986; colobus: Dunbar, 1987). In most of these cases, stress appears to be the result of increased levels of harassment and aggression.

From a purely biological point of view, there is no reason to suggest that these two factors are mutually exclusive. Indeed, it may well be that both stress and access to food simultaneously limit a female's fecundity. Dunbar & Sharman (1983), for example, were able to show that, in baboon populations, the mean birth rate was significantly influenced by both habitat quality and the adult sex ratio and that these two effects were independent of each other. This being the case, it is likely that in some populations one factor may be in operation, while in other populations both factors may be acting in concert to depress a female's fertility still further. There is some circumstantial evidence from the gelada to suggest that this may be the case: there appears to be a shift from stress to food access as the main factor suppressing fertility as altitude increases (Dunbar, 1989b).

A more serious problem is that these two mechanisms are in fact extremely difficult to separate in practice since competition for access to limited food supplies can be expected to generate a great deal of harassment and stress for low-ranking animals. Although ecologists have tended to emphasize food as the primary determinant of most biological processes, the evidence presented in most cases does not differentiate unequivocally between the two processes.

Although evidence for the third (purely behavioural) mechanism exists in other taxa (e.g. birds: Wiley, 1973); there is no evidence from any primate field studies to suggest that high-ranking females ever successfully prevent low-ranking females from mating with males (Harcourt, 1987). Although high-ranking female gelada do harass lower-ranking females more frequently when the latter come into oestrus (Dunbar, 1980b) and do tend to monopolize access to the male for at least part of the day (Dunbar, 1983), low-ranking individuals neither mate nor, more importantly, receive ejaculation from the male significantly less often than high-ranking females.

The fourth of the possible mechanisms for the reduction in fecundity in some females (namely, lactational amenorrhoea induced by the mechanical stimulation of a sucking infant) has been thoroughly documented in humans (McNeilly & McNeilly, 1979; Howie *et al.*, 1982, Gross & Eastman, 1985). These studies have demonstrated that it is the frequency rather than the duration of sucking that is instrumental in suppressing reproductive function in women. There is now evidence to suggest a similar effect in at least some species of primates (macaques: Gomendio, 1988; gorilla: Stewart, 1988) as well as in other mammals (e.g. deer: Loudon *et al.*, 1983). Other studies have indicated that females who wean their offspring early are likely to conceive again more quickly than late-weaning females (baboons: Nicholson, 1987; vervets: Whitten, 1982; Lee, 1987; macaques: Hiraiwa, 1981; Simpson *et al.*, 1981).

The details of the physiological processes involved remain unclear at present. Indeed, it is quite possible that all three hypotheses for which there is *prima facie* evidence are in fact closely related to each other. Studies of human mothers in rural communities, for example, suggest that the mother's nutritional condition influences her fertility only indirectly via the infant's higher sucking frequency. Poorly fed mothers produce little milk, which causes the infant to suck more often in order to stimulate milk production by increasing prolactin output; amenorrhoea may then be prolonged as a result of high prolactin levels. On the other hand, it may be that sucking stimulates the production of opiates as well as prolactin, so that reproductive suppression could equally well be mediated via an opiate pathway; high prolactin levels may thus only be incidentally correlated with reproductive suppression. Similarly, poor physical condition (i.e. semi-starvation) may be physically stressful, thereby triggering opiate production.

If this interpretation is correct, then other sources of stress that are independent of sucking rates may well suppress ovulatory function in the absence of nursing infants. This is certainly implied by the original experimental studies of captive ovariectomized talapoin monkeys by Bowman *et al.*, (1978) and by field data on the gelada (Dunbar 1980b, 1989b). In the latter case, the return to reproductive condition occurs some 6–12 months *after* weaning and the cessation of lactation. Moreover, although females tend to recommence cycling together, low-ranking females undergo more menstrual cycles before conceiving, suggesting that the problem lies in the suppression of ovulation rather than the return to menstrual cycling *per se*.

In conclusion, it seems clear that we are far from understanding the physiological mechanisms involved in the suppression of reproductive function in female primates. All three mechanisms for which evidence can be adduced may be independently implicated, or they may act through the same physiological pathway. In the latter case, it seems very likely that this common pathway involves endogenous opiates.

References

Abbott, D. H. (1984). Behavioural and physiological suppression of fertility in subordinate marmoset monkeys. *American Journal of Primatology*, 6, 169–86.

Abbott, D. H. (1987). Behaviourally mediated suppression of reproduction in female primates. *Journal of Zoology, London*, 213, 455–70.

Abbott, D. H. & Hearn, J. P. (1978). Physical, hormonal and behavioural aspects of sexual development in the marmoset monkey, *Callithrix jacchus*. *Journal of Reproduction and Fertility*, 53, 155–66.

Abbott, D. H., Keverne, E. B., Moore, G. F. & Yodyinguad, U. (1986). Social suppression of reproduction in subordinate talapoin monkeys, *Miopithecus talapoin*. In *Primate Ontogeny*, ed. J. Else & P. C. Lee, pp. 329–41. Cambridge: Cambridge University Press.

Altmann, J., Altmann, S. A., Hausfater, G. & McCuskey, S. A. (1977). Lifehistory of yellow baboons: physical development, reproductive parameters and infant mortality. *Primates*, 18, 315–30.

Altmann, J., Altmann, S. A. & Hausfater, G. (1988). Determinants of reproductive success in savannah baboons (*Papio cynocephalus*). In *Reproductive Success*, ed. T. H. Clutton-Brock, pp. 403–18. Chicago: Chicago University Press.

Bowman, L. A., Dilley, S. R. & Keverne, E. B. (1978). Suppression of oestrogen-induced LH surge by social subordination in talapoin monkeys. *Nature, London*, 275, 56–8.

Bulger, J. & Hamilton, W. J. (1987). Rank and density correlates of inclusive fitness measures in a natural chacma baboon (*Papio ursinus*) troop. *American Journal of Primatology*, 8, 635–50.

Caughley, G. (1977). *Analysis of Vertebrate Populations*. Chichester: Wiley.

Cheney, D. L. (1983). Extrafamilial alliances among vervet monkeys. In *Primate Social Relationships*, ed. R. A. Hinde, pp. 278–86. Oxford: Blackwell Scientific Publications.

Cheney, D. L., Seyfarth, R. M., Andelman, S. J. & Lee, P. C. (1988). Factors affecting reproductive success in vervet monkeys. In *Reproductive Success*, ed. T. H. Clutton-Brook, pp. 384–402. Chicago: Chicago University Press.

Cox, P. R. (1976). *Demography*. 5th edn. Cambridge: Cambridge University Press.

Datta, S. (1983). Relative power and the acquisition of rank. In *Primate Social Relationships*, ed. R. A. Hinde, pp. 93–103. Oxford: Blackwell Scientific Publications.

Dittus, W. P. J. (1979). The evolution of behaviours regulating density and age-specific sex ratios in a primate population. *Behaviour*, 69, 265–302.

Dittus, W. P. J. (1986). Sex differences in fitness following a group take-over among toque macaques: testing models of social evolution. *Behavioural Ecology and Sociobiology*, 19, 257–66.

Dolhinow, P. C., McKenna, J. & von der Haar Laws, J. (1979). Rank and reproduction among female langur monkeys: age and improvement (they're not just getting older, they're getting better). *Journal of Aggressive Behaviour*, 5, 19–30.

Drickamer, L. C. (1974). A ten-year summary of reproductive data for free-ranging *Macaca mulatta*. *Folia Primatologica*, 21, 61–80.

Dunbar, R. I. M. (1980a). Demographic and lifehistory variables of a population of gelada baboons (*Theropithecus gelada*). *Journal of Animal Ecology*, 49, 485–506.

Dunbar, R. I. M. (1980b). Determinants and evolutionary consequences of dominance among female gelada baboons. *Behavioural Ecology and Sociobiology*, 7, 253–65.

Dunbar, R. I. M. (1983). Structure of gelada baboon reproductive units. IV. Integration at group level. *Zeitschrift fur Tierpsychologie*, 63, 265–82.

Dunbar, R. I. M. (1984). *Reproductive Decisions: An Economic Analysis of Gelada Baboon Social Strategies*. Princeton: Princeton University Press.

Dunbar, R. I. M. (1987). Habitat quality, population dynamics and group composition in colobus monkeys (*Colobus guereza*). *International Journal of Primatology*, 8, 299–330.

Dunbar, R. I. M. (1988). *Primate Social Systems*. London: Chapman & Hall.

Dunbar, R. I. M. (1989a). Social systems as optimal strategy sets: the costs and benefits of sociality. In *Comparative Socioecology*, ed. V. Standen & R. Foley, pp. 131–149. Oxford: Blackwell Scientific Publications.

Dunbar, R. I. M. (1989b). Reproductive strategies of female gelada baboons. In *Sociobiology of Sexual and Reproductive Strategies*, ed. A. Rasa, C. Vogel & E. Voland, pp. 74–92. London: Chapman & Hall.

Dunbar, R. I. M. & Dunbar, P. (1976). Contrasts in social structure among black-and-white colobus monkey groups. *Animal Behaviour*, 24, 84–92.

Dunbar, R. I. M. & Sharman, M. (1983). Female competition for access to males affects birth rate in baboons. *Behavioural Ecology and Sociobiology*, 13, 157–9.

Fa, J. E. (1986). *Use of Time and Resources by Provisioned Troops of Monkeys*. Basel: Karger.

Fairbanks, L. A. & McGuire, M. T. (1984). Determinants of fecundity and reproductive success in captive vervets. *American Journal of Primatology*, 7, 27–38.

Fairbanks, L. A. & McGuire, M. T. (1986). Age, reproductive value and dominance related behaviour in vervet monkey females: cross-generational influences on social relationships and reproduction. *Animal Behaviour*, 34, 1710–21.

Frisch, R. E. (1978). Nutrition, fatness and fertility; the effect of food intake on reproductive ability. In *Nutrition and Human Reproduction*, ed. W. H. Mosley, pp. 91–122. New York: Plenum Press.

Frisch, R. E. (1982). Malnutrition and fertility. *Science*, 215, 1272–3.

Garber, P. A., Moya, L. & Malaga, C. (1984). A preliminary study of the moustached tamarin monkey (*Sanguinus mystax*) in northeastern Peru:

questions concerned with the evolution of a communal breeding system. *Folia Primatologica*, 42, 17–33.

Gautier-Hion, A. & Gautier, J. P. (1976). Croissance, maturité sexuelle et sociale reproduction chez les cercopithecines forestiers africaine. *Folia Primatologica*, 26, 165–84.

Goldizen, A. W. (1987). Tamarins and marmosets: communal care of offspring. In *Primate Societies*, ed. B. Smuts, D. Cheney, R. Seyfarth, R. Wrangham & T. Struhsaker, pp. 34–43. Chicago: Chicago University Press.

Gomendio, M. (1988). Mother–Offspring Relationships and Consequences for Fertility in Rhesus Macaques. Ph.D. thesis, University of Cambridge.

Goodall, J. (1983). Population dynamics during a 15-year period in one community of free-living chimpanzees in the Gombe National Park. *Zeitschrift für Tierpsychologie*, 61, 1–60.

Gouzoules, H., Gouzoules, S. & Fedigan, L. (1982). Behavioural dominance and reproductive success in female Japanese monkeys (*M. fuscata*). *Animal Behaviour*, 30, 1138–50.

Gross, B. A. & Eastman, C. J. (1985). Prolactin and the return of ovulation in breast-feeding women. *Journal of Biosocial Science, Supplement*, 9, 25–42.

Hadidian, J. & Bernstein, I. S. (1979). Female reproductive cycles and birth data from an Old World monkey colony. *Primates*, 20, 429–42.

Harcourt, A. H. (1987). Dominance and fertility among female primates. *Journal of Zoology, London*, 213, 471–87.

Hiraiwa, M. (1981). Maternal and alloparental care in a troop of free-ranging Japanese monkeys. *Primates*, 22, 309–29.

Howie, P. W., McNeilly, A. S., Houston, A. S., Cook, A. & Boyle, H. (1982). Fertility after childbirth: postpartum ovulation and menstruation in bottle and breast feeding mothers. *Clinical Endocrinology*, 17, 323–32.

Janson, C. (1985). Aggressive competition and individual food consumption in wild brown capuchin monkeys (*Cebus apella*). *Behavioural Ecology and Sociobiology*, 18, 125–38.

Lee, P. C. (1987). Nutrition, fertility and maternal investment in primates. *Journal of Zoology, London*, 213, 409–22.

Loudon, A. S., McNeilly, A. S. & Milne, J. A. (1983). Nutrition and lactational control of fertility in red deer. *Nature, London*, 302, 145–7.

McNeilly, A. S. & McNeilly, J. R. (1979). Effects of lactation on fertility. *British Medical Bulletin*, 35, 151–4.

Menard, N., Vallet, D. & Gautier-Hion, A. (1985). Démographie et reproduction de *Macaca sylvanus* dans différent habitats en Algérie. *Folia Primatologica*, 44, 65–81.

Mori, A. (1975). Signals found in the grooming interactions of wild Japanese monkeys of the Koshima troop. *Primates*, 16, 107–40.

Mori, A. (1979). Analysis of population changes by measurement of body weight in the Koshima troop of Japanese monkeys. *Primates*, 20, 371–97.

Nicholson, N. A. (1987). Infants, mothers and other females. In *Primate Societies*, ed. B. Smuts, D. Cheney, R. Seyfarth, R. Wrangham & T. Struhsaker, pp. 330–42. Chicago: Chicago University Press.

Ohsawa, H. & Dunbar, R. I. M. (1984). Variations in the demographic structure and dynamics of gelada baboon populations. *Behavioural Ecology and Sociobiology*, 15, 231–40.

Ricklefs, R. E. (1973). *Ecology*. London: Nelson.
Rowell, T. E. & Richards, S. M. (1979). Reproductive strategies of some African monkeys. *Journal of Mammalogy*, 60, 58–69.
Sackett, G. P., Holm, R. A., Davis, A. E. & Farhenbruck, E. E. (1975). Prematurity and low birth weight in pigtail macaques: incidence, prediction and effects on infant development. In *Proceedings of the Fifth Congress of the International Primatological Society*, ed. S. Kondo, M. Kawai, A. Ehara & S. Kawamura. Tokyo: Japan Science Press.
Sade, D. S., Cushing, K., Cushing, P., Dunaid, J., Figueroa, A., Kaplan, J., Lauer, C., Rhodes, D. & Schneider, J. (1976). Population dynamics in relation to social structure on Cayo Santiago. *Yearbook of Physical Anthropology*, 20 253–62.
Silk, J. B. (1988). Social mechanisms of population regulation in a captive group of bonnet macaques (*Macaca radiata*). *American Journal of Primatology*, 14, 111–24.
Simpson, M. J., Simpson, A., Hooley, J. & Zunz, M. (1981). Infant-related influences on birth intervals in rhesus monkeys. *Nature*, 290, 49–51.
Smith, D. G. (1982). A comparison of the demographic structure and growth of free-ranging and captive groups of rhesus monkeys (*Macaca mulatta*). *Primates*, 23, 24–30.
Smuts, B. B. (1985). *Sex and Friendship in Baboons*. New York: Aldine.
Stewart, K. J. (1988). Suckling and lactational anoestrus in wild gorillas (*Gorilla gorilla*). *Journal of Reproduction and Fertility*, 83, 627–34.
Strum, S. C. & Western, D. (1982). Variations in fecundity with age and environment in olive baboons (*Papio anubis*). *American Journal of Primatology*, 3, 61–76.
Sugiyama, Y. & Ohsawa, H. (1982). Population dynamics of Japanese monkeys with special reference to the effect of artificial feeding. *Folia Primatologica*, 39, 238–63.
Takahata, Y. (1980). The reproductive biology of a free-ranging troop of Japanese monkeys. *Primates*, 21, 303–29.
van Schaik, C. P. (1983). Why are diurnal primates living in groups? *Behaviour*, 87, 120–44.
van Schaik, C. P. & van Noordwijk, M. A. (1985). Interannual variability in fruit abundance and the reproductive seasonality of Sumatran long-tailed macaques. *Journal of Zoology, London*, 206, 533–49.
Whitten, P. L. (1982). Female Reproductive Strategies Among Vervet Monkeys. Ph.D. thesis, Harvard University.
Whitten, P. L. (1983). Diet and dominance among female vervet monkeys (*Cercopithecus aethiops*). *American Journal of Primatology*, 5, 139–59.
Wiley, R. H. (1973). Territoriality and non-random mating in sage grouse, *Centrocercus urophasianus*. *Animal Behaviour Monographs*, 6, 85–169.
Wilson, E. O. & Bossert, W. H. (1971). *A Primer of Population Biology*. Sunderland (Mass.); Sinauer.
Wilson, M. E., Gordon, T. P. & Bernstein, I. S. (1978). Timing of births and reproductive success in rhesus monkey social groups. *Journal of Medical Primatology*, 7, 202–12.
Winkler, P., Loch, H. & Vogel, C. (1984). Life history of hanuman langurs (*Presbytis entellus*): reproductive parameters, infant mortality and troop development. *Folia Primatologica*, 43, 1–23.

Wolfe, L. D. (1979). Sexual maturation among members of a transported troop of Japanese macaques (*Macaca fuscata*). *Primates*, 20, 411–18.
Wolfe, L. D. (1984). Female rank and reproductive success among Arashiyama B Japanese macaques (*Macaca fuscata*). *International Journal of Primatology*, 5, 133–43.

3 *Biological aspects of fertility among Third World populations*

LYLIANE ROSETTA

Introduction

The absence of mechanization and transport facilities in the Third World requires subsistence farmers, horticulturalists and hunter-gatherers to sustain moderate to high energy expenditure (Lawrence *et al.*, 1985; Bleiberg *et al.*, 1980; Bentley, 1985). Their diet is mainly derived from non-cultivated or home-grown food. Marked seasonal food shortage and chronic malnutrition are common occurrences (Bahuchet, 1988; Pagezy, 1984; Swinton, 1988; Bailey & Peacock, 1988; Hill *et al.*, 1984), and a varying form of adaptation might be needed to achieve a 'long-term energy equilibrium' (Ferro-Luzzi, 1988).

This is particularly relevant to rural areas where the phenomena concerning fertility are more acute than those observed in urban areas in the same countries. Recent immigrants to large urban areas have generally changed their diet to a more diversified one. At the same time women seem to reduce their physical workload, as many of them work as maids or sell home-baked food in the market, which is a quite different pattern of energy expenditure from that of the daily activity of women living in the bush. Another difference between urban and rural areas is the more widespread use of modern contraceptives in towns, including sterilization for men, or the use of long-duration contraceptives for women, such as six monthly injections of Depo-Provera.

In Third World populations, breastfeeding traditionally continues for long periods of time, and this is accompanied by a long duration of postpartum amenorrhoea with long birth intervals. Health care facilities are limited and the child mortality rate up to the age of five years is generally high. Many populations live in harsh physical environments in terms of climate or altitude. These include groups living in the polar circle, the equatorial and tropical areas, or at high altitude.

This review will consider the known biological effects of environment and lifestyle on the fertility of Third World populations. Infertility due to pathology from endocrine causes or as a consequence of infectious or parasitic diseases will not be discussed, nor will that due to social behaviours, such as sexual taboos longer than the duration of temporary infertility, long lasting migrations or the use of modern contraception.

The biological effects of the physical environment on reproductive function

Altitude

The effect of high altitude has been studied in the highlands of South America as well as in the Himalayan areas of Asia. All of these studies show that high altitude has some adverse effect on fecundity and fertility, including a suggestion of slightly delayed maturation at puberty, longer waiting time to first birth (Bangham & Sacherer, 1980), and reduced fertility (Gupta, 1980). These results were questioned at the beginning of the eighties by Goldstein *et al.* (1983), who argued that cultural and social behaviour are confounding factors affecting the exposure of females to the risk of intercourse. Goldstein *et al.* (1983) cite later age at marriage, later age at first birth and the existence of fraternal polyandry in some communities as examples of such cultural and social behaviour.

It seems, however, that more recent studies which control for socio-cultural factors show populations living at very high altitude to have significantly longer birth intervals (Laurenson *et al.*, 1985), and a trend toward later age at menarche (Malik & Hauspie, 1986). These results cannot be attributed solely to hypoxia, however, as several other factors such as low socio-economic status, low nutritional intakes, and hard physical work may also contribute to the long birth intervals observed.

The only clear effect of hypoxia seems to be its role in foetal growth retardation. Newborns at high altitude have lower birth weights, and more are born prematurely (Kashiwazaki *et al.*, 1988; Haas *et al.*, 1989). Consequently, they have an increased risk of early mortality, and this can play a role in total fertility (Dutt, 1980).

Photoperiod

The effect of photoperiodicity on sexual maturity has been well studied in ewe lambs (Foster and Yellon 1985; 1987), but there are few studies concerning human populations. A recent study of pituitary ovarian function, carried out in the northern part of Finland among endurance runners, joggers and their respective control groups, has taken into account the seasonal shift of photoperiod (Ronkainen *et al.*, 1985). The luminosity in this part of Finland has a large seasonal variation from two hours a day in December to twenty-two hours a day in June. Comparison of the concentration of serum follicle stimulating hormone (FSH), luteinizing hormone (LH), prolactin (PRL), estradiol (E2), progesterone (P) and testosterone (T) during one menstrual cycle in autumn and another one in spring shows a significant seasonal difference in ovarian hormones, without change in anterior pituitary hormone concentration. There are lowered E2, P, and T levels in autumn, and this is an exercise-independent phenomenon. The long distance runners show smaller autumn–spring differences but they do more training in spring, at the time of high luminosity, and less training in autumn, during the dimmer photoperiod time. High luminosity in the spring seems partly to overcome the inhibiting effect of strenuous exercise on reproductive endocrinology. But the slight photoperiodic effect on hormonal regulation in humans, only suspected in extreme shifts of luminosity, was not confirmed in Minnesota (USA) where there is a smaller contrast in luminosity (Sundararaj *et al.*, 1978).

Climate

Seasonality of conception is still observed in many populations in the Third World (Crittenden & Baines, 1986; Sindiga, 1987; Fergusson, 1987; Bantje, 1988; Huss-Ashmore, 1988). There is a complex interaction between marked variations in the climate and the consequences for diet, workload, marriage, child mortality and all the variables influencing periodicity of conception.

Comparing the birth patterns of two distinct populations living in two markedly different ecological zones, the Canadian Arctic and Papua New Guinea, Condon & Scaglion (1982) stressed the difficulties in delimiting the respective action of environmental changes, and social and biological rhythms in birth seasonality.

Direct effects of temperature and humidity variation on reproductive function have never been isolated and their biological action, if any, never demonstrated. In a recent study among nomadic Turkana in Kenya, Leslie & Fry (1989) mentioned that 'many of the factors that are most likely to contribute to birth seasonality in Turkana are directly related to food availability'. This suggests an indirect effect of the environment through variables such as nutrition, which is known by physiologists to influence the regulation of reproductive function in a wide range of mammals (Loudon, 1987).

The biological effects of lifestyle on fertility among Third World populations

An example from West Africa

Some years ago, being particularly interested in the relationship between chronic malnutrition and fertility in humans, I carried out a study of a rural

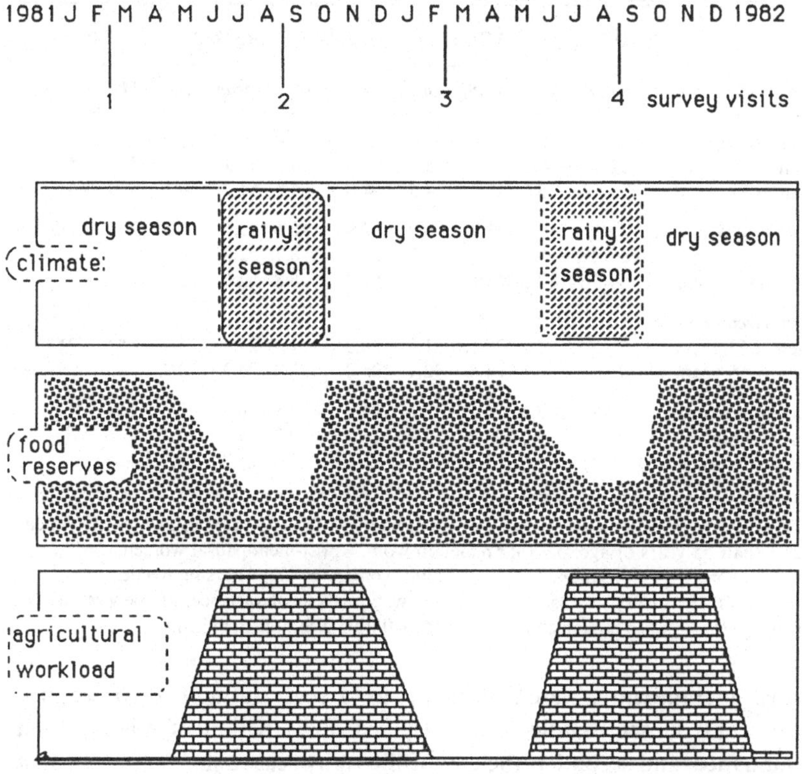

Figure 3.1. Seasonality, food reserves, agricultural workload and timetable of a fertility survey carried out among a rural population in Senegal.

population in Senegal, where the climate shows very markedly seasonality (Figure 3.1). There is a short rainy season from June to September, the rest of the year being dry. Temperatures are lowest in December and January, and highest between March and October. The more difficult period of the year for people is during the rainy season, when both temperature and humidity are high. It is a time of high incidence of parasitic diseases, mainly malaria, concurrent with possible food shortages and high workloads. People grow peanuts and millet. There is a single annual harvest at the end of the rainy season and generally the food reserves are not enough to last until the next harvest. In this case, some months before the new harvest, the head of the household will slowly reduce the daily food intake of the family, as each morning he will give those who prepare the food less millet from the household store than on previous occasions.

In order to study seasonal change, anthropometric data, fertility and

Table 3.1. *Seasonal variations in weight and arm circumference in Serere adults in Senegal (from Rosetta, 1986)*

	DS1981		*RS1981*		*DS1982*		*RS1982*[a]			
	mean	SD	*mean*	SD	*mean*	SD	*mean*	SD	*(I)*	*(II)*
Weight (kg)										
men <55 years[b]	62.4	6.1	60.3	5.9	61.6	6.3	59.9	5.2	***	*
men >55 years[c]	55.8	5.6	52.7	4.5	54.6	4.8	54.2	6.1		
p.menop women[d]	53.1	10.4	51.9	9.9	52.9	10.6	52.1	8.3	NS	NS
nn/np women[e]	55.2	8.3	53.1	6.7	53.7	8.1	52.8	6.7		
nursing women[f]	54.8	7.8	53.4	6.7	51.9	6.8	52.6	6.5		
Arm circumference (cm)										
men <55 years	28.5	2.1	27.3	1.7	28.2	1.9	27.3	1.9	***	***
men >55 years	25.5	2.5	24.6	2.1	25.2	2.4	24.9	2.2		
p.menop women	26.8	3.5	26.1	3.5	26.7	3.4	26.4	3.2	NS	NS
nn/np women	27.1	3.0	25.9	2.6	26.2	2.3	26.4	2.1		
nursing women	26.2	2.6	25.6	2.4	25.6	2.3	25.3	2.2		

[a] DS = Dry Season, RS = Rainy Season; [b] men younger than 55 years of age, n = 57; [c] men older than 55 years of age, n = 19; [d] p.menop wom. = post-menopausal women, n = 52; [e] nn/np women = neither nursing nor pregnant women, n = 23; [f] nursing women throughout the survey, n = 25; *(I)*, *(II)*: two-way analysis of variance, *(I)* between age groups, *(II)* between seasons; * $p<0.05$; *** $p<0.001$; NS: non significant.

nutritional measures were collected twice a year, during two consecutive years: first in the middle of the dry season, a period of relative food abundance, and second at the end of the rainy season just before the first annual harvest, as late as possible into the period of scarcity. This work took place in 1981 and 1982, in the rural community of Ndiaganiao, in a randomly selected sample of 40 Serere households, the dominant ethnic group in this area.

The most striking feature of sex differences in the seasonal variations of adult nutritional status is the remarkable capacity of women to cope with the adverse effects of seasonal change (Rosetta, 1986). There is no seasonal effect nor age effect in women for weight and arm circumference. In contrast Serere men show important seasonal variations and a large age effect for the same variables (Table 3.1). For the other anthropometric variables, Serere women show no significant seasonal variations. Women in this population therefore maintain sufficient energy reserves to avoid the mobilization of muscle during periods of hardship. In contrast, insufficient fat reserves lead men to mobilize protein reserves at a time of high physical workload. This is an indicator of the fragility of the nutritional balance in the population.

The group of women who breastfed throughout the study have a consistently higher mean Quetelet's index (weight/height^2) than that reported as a minimal value necessary for ovulation. 20.3 ± 1.9 is the lowest mean Quetelet index recorded in the group of lactating women, to be compared to 18.2 considered by Frisch & McArthur (1974) as the limit for the same height necessary for restoration of menstrual cycle.

The mean duration of postpartum amenorrhoea in the Serere population is 18.2 ± 0.6 months, with a range of 2 to 26 months (Rosetta, 1989). The risk of resumption of menses shows an exponential increase after only 12 months for mothers of babies still alive. Using a proportional hazard model (Cox model) to analyse the duration of postpartum amenorrhoea, eleven variables were investigated, some of them concerning the child under study (sex, season of birth, birth rank, survival status, survival duration, weaning status, breastfeeding duration), others relating to the mother (mother age at birth, number of miscarriages and number of stillbirths before the pregnancy under study, family size). Only three factors were statistically significantly associated with duration of amenorrhoea. Breastfeeding duration is positively related (regression coefficient $= +0.33$, $p < 0.001$) whilst the state of survival of the child (alive or deceased), and its feeding state (completely weaned or not) is negatively related (regression coefficient $= -1.74$, $p < 0.001$ and -0.84, $p < 0.02$ respectively).

The mean duration of breastfeeding for Serere children is 22.8 ± 0.6 months, with a range of 13 to 36 months. The only significant difference in breastfeeding duration is between the group of mothers who resumed menses at least once during the intervening month (20.9 ± 0.7 months), and the group of mothers still amenorrhoeic at the time of the last visit (25.4 ± 2.2 months). This result gives evidence of the decisive role played by the length of postpartum amenorrhoea in the determination of breastfeeding duration. The graph of child mortality before five years of age has a very marked shape (Figure 3.2). Mortality is highest in the neonatal period and at the end of weaning. However, if the death of the baby occurs after the end of the postpartum amenorrhoea of the mother, it does not influence the duration of the next birth interval (Rosetta, 1988a).

Serere people have a lifestyle similar to that of a great number of rural Third World populations, where women take part actively in the fieldwork during the rainy season. Serere women are also in charge of gathering water, wood or other fuels and preparing meals all year long. Some of these tasks require important physical effort as evaluated in other studies carried out in similar West African rural populations. Mean energy expenditure was estimated by Bleiberg *et al.* (1980) as 11.15 MJ (2320 kcal)/day during the dry season and 13.9 MJ (2890 kcal)/day during the rainy season among women farmers in Upper Volta. Another study gave a

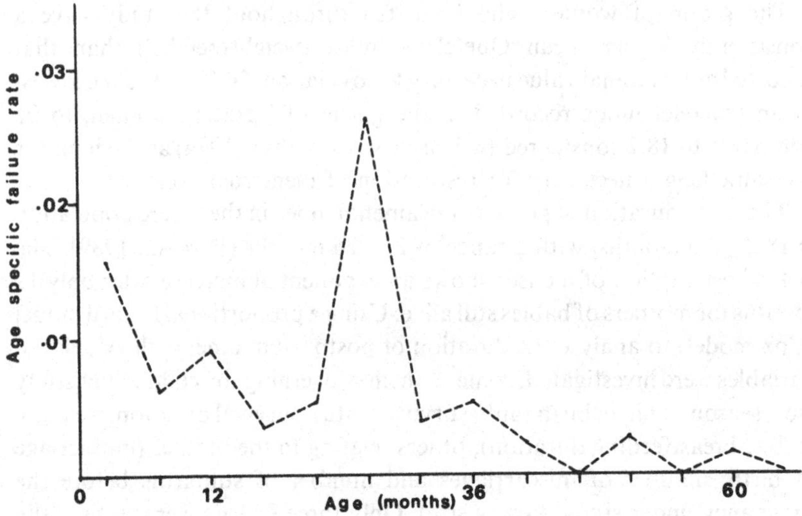

Figure 3.2. Mortality before five years of age among Serere children in Senegal. (Hazard-rate estimates also called failure rate.)

range of energy expenditure among Gambian women from 9.4 MJ (1960 kcal/day) during the dry season to 10.9 MJ (2270 kcal)/day during the rainy season (Lawrence *et al.*, 1987).

Serere food intake also has features typical of diets of other Third World populations (Rosetta, 1988b), including large quantities of carbohydrates and very small quantities of animal protein (less than 10% of total protein intake). Energy requirements are scarcely satisfied during the rainy season, specially in large households. The intakes of lipid, zinc, vitamin A, vitamin B and folate did not meet the recommendations proposed by FAO/WHO (1973) whatever the season. By inference, marginal energy and nutrient deficiences are a common occurrence among Serere people. Long periods of postpartum amenorrhoea of Serere lactating women may thus be due to the concomitant stresses of nutritional deficiencies and high energy expenditure due to hard physical work (Rosetta, 1988c).

Sedentarization, urbanization, change in behaviour and variability in the duration of postpartum amenorrhoea

A few generations ago, women in Western countries breastfed their children for a long time, and experienced a duration of post-partum amenorrhoea similar to that of women in the Third World today. Liestol *et al.* (1988) have studied the length of postpartum amenorrhoea during ongoing lactation in Norway from 1860 until the present. They found a substantial decline in this variable from 1900 onward. The median duration of postpartum amenorrhoea for women who had given birth before 1900 was twelve months, compared with six months for those women giving birth after 1900. The authors largely attributed the decline in the length of amenorrhoea to a change in breastfeeding pattern, which involved either a reduction of the intensity of suckling due to a supplementation for the baby, or reduction of night nursing due to the separation of the baby from his/her mother during the night.

The duration of postpartum amenorrhoea is related to the nursing pattern; the suckling duration (more minutes per episode), the suckling frequency more frequent day and nighttime feeds) and number of episodes per night are the most important factors for the delay in onset of post-partum menstruation (Jones, 1989). The suckling pattern alone cannot, however, explain the wide range of variability in inter-birth intervals observed in various human populations not using contraceptives (Loudon, 1987).

A change in lifestyle seems enough to modify indicators of reproductive function in the same ethnic group. The low fertility among !Kung hunter-gatherers has been attributed to heavier workload and less suitable weaning food than for !Kung women settled at Bantu cattleposts (Bentley, 1985). With sedentarization there is a trend to earlier maturation at puberty, and a shortening of the interval between two successive births. In this case the main factor interacting with fertility seems to be energy expenditure.

Another fascinating result is that obtained by Lunn and his colleagues in Gambia (1981). Dietary supplementation of lactating Gambian women reduced the duration of postpartum amenorrhoea by at least six months. Nothing but the diet changed in the lifestyle of these women. Earlier decreases in plasma prolactin levels were observed for supplemented mothers not associated with modification in the volume or composition of milk taken by the child. In addition, plasma concentrations of estradiol and progesterone 'begin to rise earlier the greater the level of supplementation' (Lunn *et al.*, 1984).

Some evidence of impaired gonadal hormone levels in different Third World populations

Other hormonal studies carried out in Third World populations have emphasized the high frequency of gonadal dysfunction in these populations. Among the !Kung San, in Botswana, normal levels of gonadotrophins and very low levels of estradiol, progesterone and testosterone were found in a group of childbearing age women (Van Der Walt *et al.*, 1978).

Salivary progesterone assay in non-contracepting Bangladeshi women of low socio-economic status showed low progesterone levels, suggesting frequent occurrence of anovulatory cycles or postovulatory luteal insufficiency (Seaton & Riad-Fahmy, 1980).

The measurement of salivary progesterone levels in women from semi-nomadic or village populations in the Ituri forest of north-eastern Zaire with primary or secondary infertility was compared to those of Western women with regular menstrual cycles (Ellison *et al.*, 1986). The Ituri women showed lower levels of progesterone during the luteal phase, a later onset of luteal progesterone secretion and a large number of anovulatory cycles. Another study carried out one year later in the same ethnic group, for normally menstruating women, during a period supposed to be of highest nutritional status (Ellison *et al.*, 1989), confirmed the previous results in this population: low ovulatory frequency, with low composite progesterone profile even if ovulatory cycles alone are considered among Lese women and compared with Boston controls (Figure 3.3).

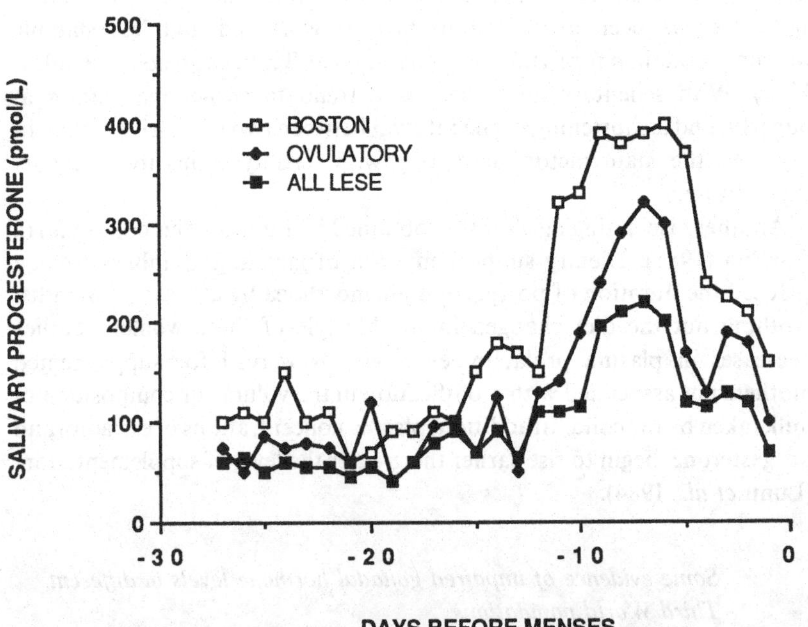

Figure 3.3. Average salivary progesterone profiles of a sample of Lese women (Zaire), the subset of ovulatory Lese cycles, and a sample of Boston women aged 23–35 years (from Ellison *et al.*, 1989, reproduced by kind permission of Peter T. Ellison).

In Papua New Guinea a survival analysis of menstrual data indicates that the median cycle length of Gainj women is approximately 40% longer on average than those of US women matched with them for gynaecological age (Johnson *et al.*, 1987). In a previous study of the same population (Wood *et al.*, 1985), Gainj women appeared to have a low reproductive output with late menarche, long intervals between births (42 months), long duration of breastfeeding (38.6 months for children surviving to 72 months of age), and early cessation of reproduction (by the age of 40 years).

Few hormonal studies have been carried out among rural populations in the Third World but those that exist provide some evidence of impaired gonadal hormone level in women, even though the level of gonadotrophins seems normal.

Is postpartum amenorrhoea just a secondary amenorrhoea?

Lifestyle and secondary amenorrhoea in Western countries

Known lifestyle features liable to cause menstrual disturbances in Western countries are given in Table 3.2.

The effects of strenuous physical exercise during adolescence on the reproductive function in girls have already been pointed out (Malina, 1983). Considerable delay in menarche has been found in ballet dancers (15.4 ± 1.9 years) compared with age-matched controls (12.5 ± 1.2 years) and music students (12.9 ± 1.9 years). This delay is sometimes associated with marked delay in skeletal maturation as assessed by bone age (Warren, 1980).

During follow-up studies, a high level of secondary amenorrhoea has been reported in the group of postmenarcheal ballet dancers with no change in body weight or the calculated percent of body fat (Warren, 1980; Abraham *et al.*, 1982).

The incidence of reproductive defects, including delayed menarche, shortened luteal phase and amenorrhoea, is higher in women who practise sports of endurance or sports demanding high energy expenditure (Toriola & Mathur, 1986). The degree and incidence of menstrual dysfunction has been reported to be directly proportional to the severity of physical exertion (Dale *et al.*, 1979) but these troubles no longer persist two to six months after cessation of training (Stager *et al.*, 1984).

Other authors consider loss of weight or low energy reserves as a causal factor in menstrual dysfunction (Frisch & McArthur, 1974).

Nutritional status and diet can influence oestrogen metabolism. The effects of vegetarianism have been suspected in the development of infertility in women. Vegetarian diets increase faecal excretion of oestrogen

Table 3.2. *Factors in the lifestyle of Western women suspected to play a role in menstrual dysfunction*

Strenuous exercises	Warren, 1980
	Abraham *et al.*, 1982
	Toriola and Mathur, 1986
	Dale *et al.*, 1979
	Stager *et al.*, 1984
Loss of weight	Frisch & McArthur, 1974
Vegetarian diet	Goldin *et al.*, 1982
	Pirke *et al.*, 1986
Undernutrition	Loudon, 1987
	Warren, 1983
	Kirkwood *et al.*, 1987
	Steiner, 1987
Stress	Carr *et al.*, 1981
Breastfeeding	Short, 1983

hormones leading to a decreased plasma concentration of oestrone and oestradiol in women (Goldin *et al.*, 1982).

Stress which triggers off the hypothalamic release of endogenous opioid peptides (beta-endorphin), can modulate the secretion of gonadotrophin releasing hormone (GnRH) and the subsequent production of follicle stimulating hormone (FSH) and luteinizing hormone (LH) (Ferin, 1987).

Breatfeeding contributes, given a certain pattern of feeding on demand with frequent suckling stimuli and nighttime feeds, to a substantial increase in the duration of postpartum amenorrhoea (Short, 1983).

All these studies show that most of these factors operate together. Sportswomen with menstrual troubles are those with lowest body weight and fat, and those involved in the most strenuous physical activity. The incidence of secondary amenorrhoea is higher in marathon runners who are vegetarian than in marathon runners who are omnivorous. Acute exercise causes a rise in plasma level of beta-endorphins and training augments this effect (Carr *et al.*, 1981).

Physiological findings concerning menstrual regulation

Knobil (1980) has demonstrated that if the pulsatility of the hypothalmic generator of GnRH has stopped or occurs with inadequate frequency, the pituitary production of FSH and LH is impaired. Moreover, if oestradiol production is not adequate, a postponement of the positive feedback loop occurs, by which rising oestrogen secretion from the ovaries produces a

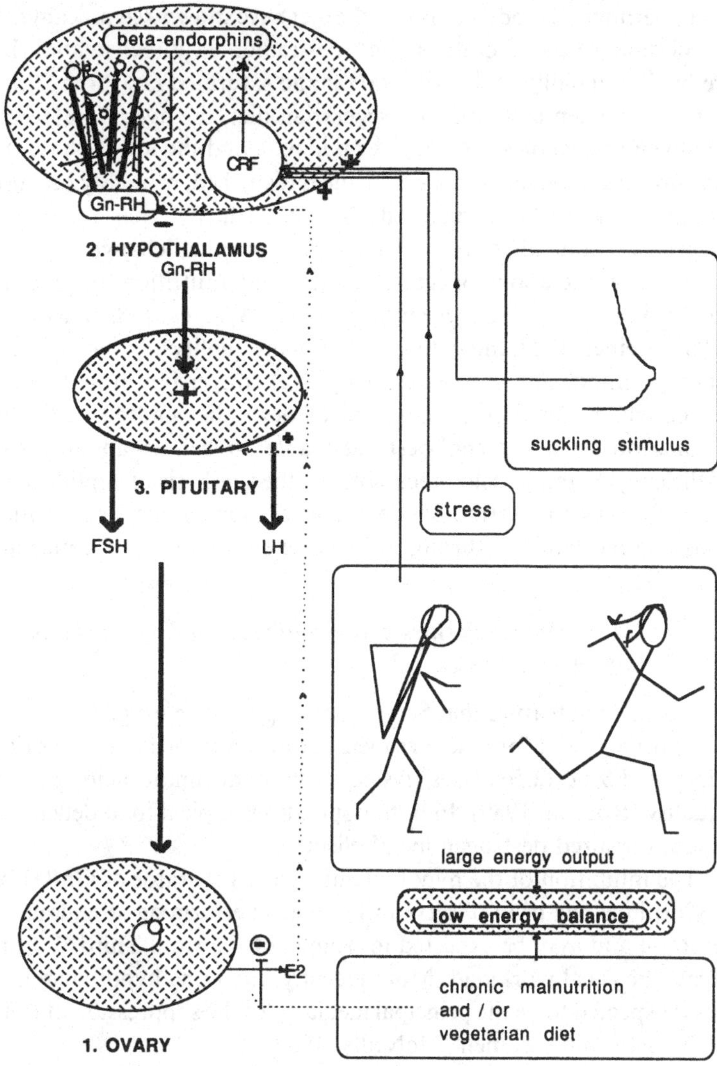

Figure 3.4. Lifestyle characteristics of women capable of affecting reproductive function at central and peripheral levels. CRF = corticotropin-releasing factor; Gn-Rh = gonadotrophin-releasing hormone; FSH = follicle stimulating hormone; LH = luteinizing hormone; E2 = oestradiol; Beta-endorphins = endogenous opioid peptides.

reflex discharge of LH that normally results in ovulation (Figure 3.4). The basic roles of a properly functioning ovary and the production of oestradiol in the regulation of the menstrual cycle have been stressed, while the hypothalamic pulse generator appears to only have a permissive effect on the production of gonadotrophins.

Experimental modifications of the diet have shown a significant reduction in plasma levels of oestrone and oestradiol, in direct relationship with reduction in daily intakes of total fat and saturated fat (Pirke *et al.*, 1986). Among women assigned to a vegetarian diet, the average LH values are significantly decreased during the midcycle and the luteal phase. Further, the process of luteinizing the follicule is disturbed, probably because of the inappropriate LH secretion during and shortly after midcycle. Non-vegetarian controls show no significant hormonal change.

Possible metabolic mechanisms of undernutrition on reproductive performance have been reviewed elsewhere (Warren, 1983; Kirkwood *et al*, 1987; Steiner, 1987) indicating that both the quality and the quantity of the diet are important. Decreased peripheral conversion of androstenedione to oestrone, impaired metabolism of oestradiol ending in the eventual formation of a catechol-oestrogen able to act as an anti-oestrogen, deficiency of amino acid precursors for the synthesis of peptide hormones, and changes in oestradiol metabolic clearance rate are mentioned as possible mechanisms through which the effects of diet are mediated.

Are there similarities between athletes and Third World amenorrhoeic women?

It is a striking feature that Serere food supply is an exaggerated version of the profile found for amenorrhoeic marathon runners: low caloric intake, lipid and zinc deficiencies, lysine as limitant amino acid, poor protein quality (Rosetta, 1989). In both populations typical food deficiencies may cause impaired oestrogen metabolism.

The inhibition of the hypothalamus pulse generator of GnRH by beta-endorphins has been observed in athletes performing strenuous physical exercise and may be expected in women of the rural Third World with a heavy physical workload. More recently, the same hypothalamic process was suspected to be the principal mediator of the suppression of LH release in breastfeeding women (McNeilly, 1988).

Conclusions

A general feature in Third World populations is chronic malnutrition with specific nutrient deficiencies in the diet. In rural areas women do physical work that occasionally requires high energy expenditure. They usually breastfeed on demand during long periods.

These general characteristics of the lifestyle of Third World populations may affect the menstrual regulation at central and peripheral levels (Figure 3.4).

At the central level, the production of beta-endorphins during the suckling stimulus, and an additional production of endogenous opioid peptides during very harsh physical work can block the hypothalamic pulsatile generator of GnRH in women of the Third World as well as in amenorrhoeic marathon runners.

At a peripheral level, chronic malnutrition and a vegetarian diet may contribute to impaired oestrogen production and, consequently, the pituitary positive feedback of oestradiol may be reduced.

The combination of the two processes is certainly necessary for long lasting amenorrhoea, and energy balance at low levels of intake may act as a switch between amenorrhoea and normal menstrual cycling. Such switching remains to be demonstrated, but if present may help to explain the variability in amenorrhoea observed in different environmental contexts.

References

Abraham, S. F., Beumont, P. J. V., Fraser, I. S. & Llewellyn-Jones, D. (1982). Body weight, exercise and menstrual status among ballet dancers in training. *British Journal of Obstetrics and Gynaecology*, **89**, 507–10.

Bahuchet, S. (1988). Food supply uncertainty among the Aka Pygmies (Lobaye, Central African Republic). In *Coping with Uncertainty in Food Supply*, ed. I. de Garine and G. A. Harrison, pp. 118–49, Oxford: Clarendon Press.

Bailey, R. C. & Peacock, N. R. (1988). Efe Pygmies of northeast Zaire: subsistence strategies in the Ituri forest. In *Coping with Uncertainty in Food Supply*, ed. I. de Garine and G. A. Harrison, pp. 88–117, Oxford: Clarendon Press.

Bangham, C. R. M. & Sacherer, J. M. (1980). Fertility of Nepalese Sherpas at moderate altitudes: comparison with high-altitude data. *Annals of Human Biology*, **7** (4), 323–30.

Bantje, H. F. W. (1988). Female stress and birth seasonality in Tanzania. *Journal of Biosocial Science*, **20** (2), 195–202.

Bentley, G. (1985). Hunter-gatherer energetics and fertility: a reassessment of the !Kung San. *Human Ecology*, **13** (1), 79–109.

Bleiberg, F. M., Brun, T. A. & Goihman, S. (1980). Duration of activities and energy expenditure of female farmers in dry and rainy seasons in Upper-Volta. *British Journal of Nutrition*, **45**, 67–75.

Carr, D. B., Bullen, B. A., Skrinar, G. S., Arnold, M. A., Rosenblatt, M., Beitins, I. Z., Martin, J. B. & McArthur, J. W. (1981). Physical conditioning facilitates the exercise-induced secretion of beta-endorphin and beta-lipotropin in women. *New England Journal of Medicine*, **305** (10), 560–2.

Condon, R. G. & Scaglion, R. (1982). The ecology of human birth seasonality. *Human Ecology*, **10** (4), 495–511.

Crittenden, R. & Baines, J. (1986). The seasonal factors influencing child malnutrition on the Nembi Plateau, Papua New Guinea. *Human Ecology*, **14** (2), 191–223.

Dale, E., Gerlach, D. H. & Wilwhite, A. L. (1979). Menstrual dysfunction in distance runners. *Obstetrics and Gynecology*, **54** (1), 47–53.

Dutt, J. S. (1980). Altitude and fertility: the confounding effect of childhood mortality – a Bolivian example. *Social Biology*, **27** (2): 101–13.

Ellison, P. T., Peacock, N. R. & Lager, C. (1986). Salivary progesterone and luteal function in two low-fertility populations of Northeast Zaire. *Human Biology*, **58** (4), 473–83.

Ellison, P. T., Peacock, N. R. & Lager, C. (1989). Ecology and ovarian function among Lese women of the Ituri forest, Zaire. *American Journal of Physical Anthropology*, **78**, 519–26.

FAO/WHO (1973). *Energy and Protein Requirements*. FAO Nutrition Meeting, Report series No. 52, Food and Agriculture Organization, Rome.

Ferro-Luzzi, A. (1988). Marginal energy malnutrition: some speculations on energy sparing mechanisms. In *Capacity for Work in the Tropics*, ed. K. J. Collins and D. F. Roberts, pp. 141–64, Cambridge: Cambridge University Press.

Ferin, M. (1987). A role for the endogenous opioid peptides in the regulation of gonadotropin secretion in the Primate. *Hormone Research*, **28**: 119–25.

Fergusson, A. G. (1987). Some aspects of birth seasonality in Kenya. *Social Sciences and Medicine*, **25** (7), 793–801.

Foster, D. L. & Yellon, S. M. (1985). Photoperiod time measurement is maintained in undernourished lambs with delayed puberty. *Journal of Reproduction and Fertility*, **75**, 203–8.

Foster, D. L. & Yellon, S. M. (1987). Absence of an increase in gonad-independent drive to pulsatile luteinizing hormone secretion during photoperiod-induced puberty. *Biology of Reproduction*, **37**, 634–9.

Frisch, R. E. & McArthur, J. W. (1974). Menstrual cycles: fatness as a determinant of minimum weight for height necessary for their maintenance or onset. *Science*, **185**, 949–51.

Goldin, B. R., Adlercreutz, H., Gorbach, S. L., Warram, J. H., Dwyer, J. T., Swenson, L. & Woods, M. N. (1982). Estrogen excretion patterns and plasma levels in vegetarian and omnivorous women. *New England Journal of Medicine*, **307**, 1542–7.

Goldstein, M. C., Tsarong, P. & Beall, C. M. (1983). High altitude hypoxia, culture, and human fecundity/fertility: a comparative study. *American Anthropologist*, **85**, 28–49.

Gupta, R. (1980). Altitude and demography among the Sherpas. *Journal of Biosocial Science*, **12**, 103–14.

Hass, J. D. Conlisk, E. & Frongillo, E. A. (1989). Fetal growth and neonatal mortality at high and low altitudes in Bolivia. *American Journal of Physical Anthropology*, **78** (2): 233.

Hill, K., Hawkes, K. Hurtado, M. & Kaplan, H. (1984). Seasonal variance in the diet of Ache hunter-gatherers in eastern Paraguay. *Human Ecology*, **12** (2), 101–35.

Huss-Ashmore, R. (1988). Seasonal patterns of birth and conception in rural highland Lesotho. *Human Biology*, **60** (3), 493–506.

Johnson, P. L., Wood, J. W., Campbell, K. L. & Maslar, I. A. (1987). Long ovarian cycles in women of Highland New Guinea. *Human Biology*, **59** (5), 837–45.

Jones, R. E. (1989). Breast-feeding and post-partum amenorrhoea in Indonesia. *Journal of Biosocial Science*, **21**, 83–100.

Kashiwazaki, H., Suzuki, T. & Takemoto, T-I. (1988). Altitude and reproduction of the Japanese in Bolivia. *Human Biology*, **60** (6), 831–45.

Kirkwood, R. N., Cumming, D. C. & Aherne, F. X. (1987). Nutrition and puberty in the female. *Proceedings of the Nutrition Society*, **46**, 177–92.

Knobil, E. (1980). The neuroendocrine control of the menstrual cycle. *Recent Progress in Hormone Research*, **36**, 53–88.

Laurenson, I. F., Benton, M. A., Bishop, A. J. & Mascie-Taylor, C. G. N. (1985). Fertility at low and high altitude in Central Nepal. *Social Biology*, **32** (1–2), 65–70.

Lawrence, M., Singh, J., Lawrence, F. & Whitehead, R. G. (1985). The energy cost of common daily activities in African women: increased expenditure in pregnancy? *American Journal of Clinical Nutrition*, **42**, 753–63.

Lawrence, M., Coward, W. A., Lawrence, F., Cole, T. J. & Whitehead, R. G. (1987). Energy requirements of pregnancy in the Gambia. *Lancet*, **ii**, 1072–6.

Leslie, P. W. & Fry, P. (1989). Extreme seasonality of births among nomadic Turkana pastoralists. *American Journal of Physical Anthropology*, **79** (1), 103–15.

Liestol, K., Rosenberg, M. & Walloe, L. (1988). Lactation and post-partum amenorrhoea: a study based on data from three Norvegian cities 1860–1964. *Journal of Biosocial Science*, **20**, 423–34.

Loudon, A. (1987). Nutritional effects on puberty and lactational infertility in mammals: some interspecies considerations. *Proceedings of the Nutrition Society*, **46**, 203–16.

Lunn, P. G., Watkinson, M., Prentice, A. M., Morell, P., Austin, S. & Whitehead, R. G. (1981). Maternal nutrition and lactational amenorrhoea. *Lancet*, **i**, 1428–9.

Lunn, P. G., Austin, S., Prentice, A. M. and Whitehead, R. G., (1984). The effect of improved nutrition on plasma prolactin concentrations and postpartum infertility in lactating Gambian women. *American Journal of Clinical Nutrition*, **39**, 227–35.

Malina, R. M. (1983). Menarche in athletes: a synthesis and hypothesis. *Annals of Human Biology*, **10** (1), 1–24.

McNeilly, A. S., (1988). Suckling and the control of gonadotrophin secretion. In *The Physiology of Reproduction*, ed. E. Knobil and J. Neil *et al.*, pp. 2323–49, New York: Raven Press Ltd.

Malik, S. L. & Hauspie, R. C. (1986). Age at menarche among high altitude bods of Ladakh (India). *Human Biology*, **58** (4), 541–8.

Pagezy, H. (1984). Seasonal hunger as experienced by the Oto and the Twa women of a Ntomba village in the Equatorial forest (lake Tumba, Zaire). *Ecology of Food and Nutrition*, **15**, 13–27.

Pirke, K. M., Schweiger, U., Laessle, R., Dickhaut, B., Schweiger, M. & Waechtler, M. (1986). Dieting influences the menstrual cycle: vegetarian versus nonvegetarian diet. *Fertility and Sterility*, **46** (6), 1083–8.

Ronkainen, H., Pakarinen, A., Kirkinen, P. & Kauppila, A. (1985). Physical exercise-induced changes and season-associated differences in the pituitary-ovarian function of runners and joggers. *Journal of Clinical Endocrinology and Metabolism*, **60** (3), 416–22.

Rosetta, L. (1986). Sex differences in seasonal variations of the nutritional status of Serere adults in Senegal. *Ecology of Food and Nutrition*, **18** (3), 231–44.

Rosetta, L. (1988a). Weaning mortality in Serere children in Senegal. *Bulletins et Mémoires de la Société d'Anthropologie de Paris*, **5** (1–2), 117–22.

Rosetta, L. (1988b). Seasonal variations in food consumption by Serere families in Senegal. *Ecology of Food and Nutrition*, **20** (4), 275–86.

Rosetta, L. (1988c). Contribution à l'étude des effets de la malnutrition chronique sur la fonction de reproduction dans l'espèce humaine. A partir d'une étude réalisée dans une population d'agriculteurs sédentaires du Sénégal. Thèse de Doctorat d'Etat ès-Sciences. Université Paris 7, France.

Rosetta, L. (1989). Breastfeeding and post-partum amenorrhoea in Serere women in Senegal. *Annals of Human Biology*, **16** (4), 311–20.

Seaton, B. & Riad-Fahmi, D. (1980). Use of salivary progesterone assays to monitor menstrual cycles in Bangladeshi women. *Journal of Endocrinology*, **87**, 21.

Short, R. V. (1983). The biological basis for the contraceptive effects of breast-feeding. *Advances in International and Child Health Care*, **3**, 27–39.

Sindiga, I. (1987). Fertility control and population growth among the Maasai. *Human Ecology*, **15** (1), 53–66.

Stager, J. M., Ritchie-Flanagan, B. & Robertshaw, D. (1984). Reversibility of amenorrhea in athletes. *New England Journal of Medicine*, **310** (1), 51–2.

Steiner, R. A. (1987). Nutritional and metabolic factors in the regulation of reproductive hormone secretion in the primate. *Proceedings of the Nutrition Society*, **46**, 159–75.

Sundararaj, N., Chern, M., Gatewood, L. Hickman, L. & McHugh, R. (1978). Seasonal behavior of human menstrual cycles: a biometric investigation. *Human Biology*, **50** (1), 15–31.

Swinton, S. M. (1988). Drought survival tactics of subsistence farmers in Niger. *Human Ecology*, **16** (2), 123–44.

Toriola, A. L. & Mathur, D. N. (1986). Menstrual dysfunction in Nigerian athletes. *British Journal of Obstetrics and Gynaecology*, **93**, 979–85.

Van der Walt, L. A., Wilmsen, E. N. & Jenkins, T. (1978). Unusual sex hormone pattern among desert-dwelling hunter-gatherers. *Journal of Clinical Endocrinology and Metabolism*, **46** (4), 658–63.

Warren, M. P. (1980). The effects of exercise on pubertal progression and reproductive function in girls, *Journal of Clinical Endocrinology and Metabolism*, **51** (3), 1150–7.

Warren, M. P. (1983). Effects of undernutrition on reproductive function in the human, *Endocrine Reviews*, **4** (4), 363–77.

Wood, J. W., Johnson, P. L. & Campbell, K. L. (1985). Demographic and endocrinological aspects of low natural fertility in highland New Guinea, *Journal of Biosocial Science*, **17**, 57–79.

4 *A preliminary report on fertility and socio-economic changes in two Papua New Guinea communities*

TUKUTAU TAUFA, VUI MEA AND JOHN LOURIE

The Wopkaimin people of the Star mountain census division were first contacted by the Australian government authorities in 1963 (Jackson, 1982). They number about 800 people (National Statistical Office, 1983) who live in seven villages scattered, over an area of about 1700 square kilometres, in the north-west corner of the Western Province of Papua New Guinea (PNG) (Figure 4.1).

Kennecott geologists, discovered copper in the area in late 1968 and by 1971 a mining camp with the first school and aid post for the whole area was built at Tabubil. The first airstrip was completed in 1972. In 1981 after the Papua New Guinea Government approved the development of a copper and gold mine, the quiet Tabubil camp with 12 buildings rapidly became a township accommodating more than 3000 workers. Mountain sites were blasted, mining facilities constructed and roads bulldozed through with the most modern technology, engineering and machinery. The first and only government aid post was built in 1982. A road connection to the Fly River port of Kiunga, 150 km to the south, was completed in 1983.

The new development pushed the Wopkaimin from the Stone Age into the modern world. Their changed environment and lifestyle had an irreversible effect on their physical, mental and spiritual wellbeing. Villages were moved to be nearer the new developments (Figure 4.1). The people left their thatched houses and built themselves shacks in their new villages out of timber, iron and plastic left over by the mining company. Men abandoned gourds for trousers, women their grass skirts for Western dress. The men have abandoned game hunting for readily available store-bought tinned fish and meat. The women do less gardening now, since rice, flour and biscuits are also readily available. In 1982 only a few young men spoke Pidgin. Now they all speak *tok pisin* (Melanesian Pidgin).

The Mt. Obree census division people of the Rigo district, Central Province, who number about 1000 (Shadlow, 1982) live in an isolated rugged mountainous area in the foothills of the Owen Stanley Range

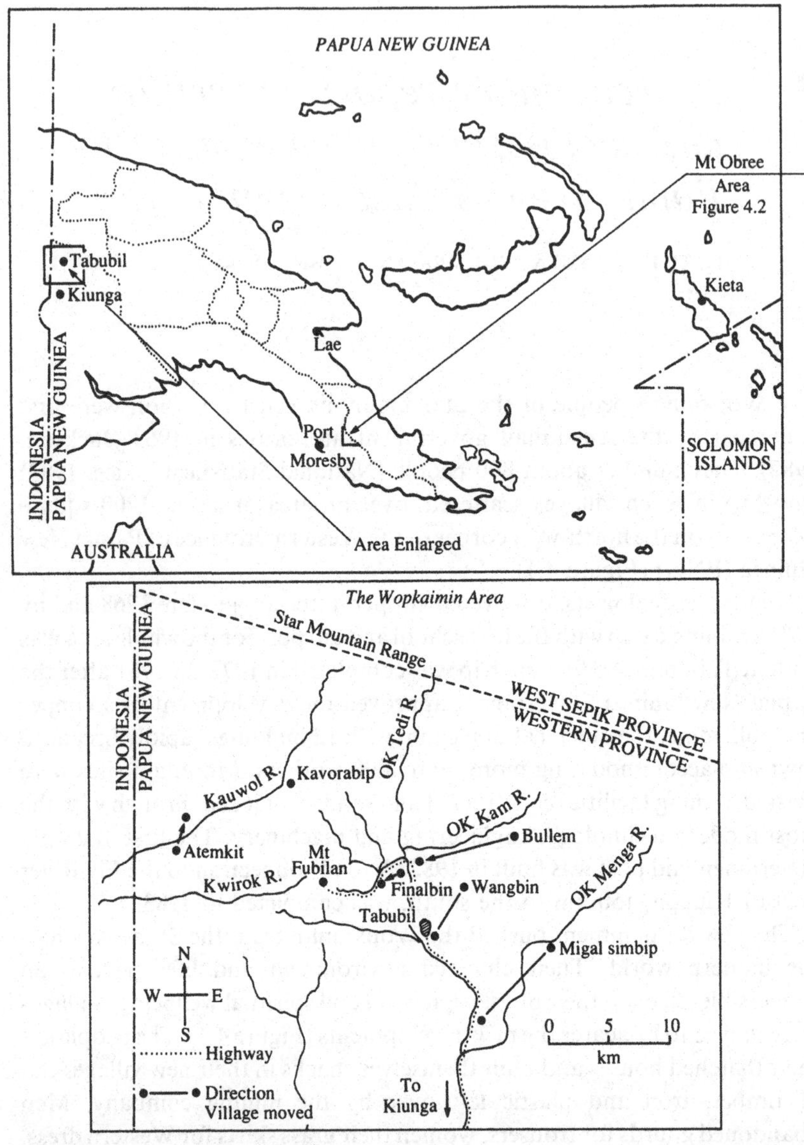

Figure 4.1. The Wopkaimin area.

(Figure 4.2). This is the 'control' group. They have similar altitude, climate, vegetation and diet. They were first contacted about 1895 when the London Missionary Society (LMS) built their headquarters at Boku in the inland Rigo district (Schlencker, 1895). The area is about two days walk from Boku or Sogeri rubber plantations (Figure 4.2). The first LMS pastor, who

Figure 4.2. Mount Obree census division Rigo district, Central Province.

started a vernacular school, was posted in the area in 1956 (personal communication, Sir Ravu Henao, retired United Church bishop, Barakau Village, Central Province, Papua New Guinea) and the first aid post orderly posted there in 1957 (personal communication, Mr Kitone, Councillor, Dorobisoro Village, Mt. Obree, Central Province, Papua New Guinea). Their first airstrip was opened in 1977 and their first government community school opened in 1978. There is no road link to the coast and economically it is a backwater of Papua New Guinea. They all speak 'Police motu'.

Method

The Wopkaimin were first visited by the authors in 1982 and the Mt. Obree people in 1983. Both areas are visited at least twice a year. The Mt. Obree work takes about two weeks as the visitors have to carry all their supplies and walk through mountainous areas. The lighter Ok Tedi component follows this immediately and is done in a week using motor vehicle and helicopter for transportation. In each village, the villagers are interviewed, clinically examined, specimens collected and first aid treatment given to the sick. The under five year olds and other eligible groups are given the appropriate standard Health Department vaccines. The authors have done this work using the same instruments over the last six years. Wide ranging health indices are monitored and only fertility aspects will be reviewed here. A village birth and death registry book is left in each village with a village recorder and this is checked during each visit.

Results

Crude birth rate

The Wopkaimin crude birth rate (CBR) in 1982 was below Papua New Guinea's average 34.2/1000 population. By 1984 this had increased to 74/1000 population, a very high CBR by world standard (United Nations World Population Chart, 1985). This phenomenal increase is related to the sudden change of their physical, social and cultural environment.

Villages were moved nearer to the new developments in 1981/82 (Figure 4.1). Those who moved were mainly the young fertile people who were employed by the Mining Company. The old were left to maintain their culture and traditions in the old villages. In the new villages the lifestyle is different. They no longer depend upon the advice of their elders for gardening, hunting and how to raise families. The last initiation ceremony was held in December 1982 and the 'house tamparan' (a traditional house where cultural items and spirits are housed) is now rotting and falling to pieces. Western food is now easily available from trade stores.

The young are the bread-winners now and they liaise with the Company and Government Officials and control the royalty and lease payment money and investments. In the old villages men sleep by themselves in the men's house while women and children sleep by themselves in their family houses. In the new village there is no men's house and the young men sleep with their wives in their own houses, an arrangement unheard of in their traditional villages. This has contributed to the explosive crude birth rate since 1984.

The Mt. Obree people on the other hand have been traditionally living with their wives in their own houses. They have also not experienced any socio-economic changes and the traditional elders' leadership is still intact. The CBR therefore has remained relatively high but stable.

Birth interval

Among the Wopkaimin the traditional birth interval obtained from limited information in 1982 was about five years. The women between births have

Table 4.1. *Crude birth rates (CBR)*[a] *in the two areas*

	Year						
Area	*1982*	*1983*	*1984*	*1985*	*1986*	*1987*	*1988*
Wopkaimin	32	33	74	79.4	82.4	78	69
Mt. Obree		45.8	51	44.8	46.6	53.7	44

[a] Annual births per 1000 population.
The Wopkaimin CBR increased substantially in 1984 and peaked in 1986 with a slight decrease later. The Mt. Obree CBR has remained relatively constant.

Table 4.2. *Mean birth intervals in months in the two areas*

	Year		
Area	*Before 1982*	*1984*	*1988*
Wopkaimin	>5 years	35 months	29.7 months
Mt. Obree		37.9 months	36 months

The mean birth intervals among Wopkaimin in 1988 is significantly less than that of 1984 (z=2.086; p=0.045).
There is no significant difference in the mean birth intervals among the Mt. Obree women between 1984 and 1988 (z=0.728; p=0.48).

Table 4.3. *Preceding birth intervals among women in the two areas*

Birth intervals in months	*Wopkaimin women*	*Mt. Obree women*
13–24	41	9
25–36	39	31
37–48	11	17
49–60	9	5
61–72	1	3

The mode among the Wopkaimin women is in the 13–24 months birth interval whereas with the Mt. Obree women the mode is in the 25–36 months birth intervals.

to make the gardens and tend the pigs. In 1984, with accurate dates of birth available, we found that the mean birth interval of two previous pregnancies was 35 months (Taufa *et al.*, 1986) (Table 4.2). In 1988 this has further decreased significantly to 29.7 months ($z=2.086$; $p=0.045$). The Mt. Obree women experienced no significant mean birth interval changes between 1984 and 1988 (Table 4.2).

Table 4.4. *Infant mortality rate (IMR)*[a] *in the two areas*

	Year						
Area	*1982*	*1983*	*1984*	*1985*	*1986*	*1987*	*1988*
Wopkaimin	52.6	371.4	90.9	100	63.4	60	57
Mt. Obree		105	191	71	136	98	92

[a] Deaths in the first year of life per 1000 live births.
Apart from the measles epidemic among the Wopkaimin in 1983, the IMR is decreasing. There was also a measles epidemic in the Mt. Obree area in 1984 but otherwise the IMR remains relatively the same.

Table 4.3 shows that the mode of birth intervals among the Wopkaimin is two years and less. The data for the Mt. Obree women on the other hand show a mode between 25–36 months. In a survey of urban villages in Port Moresby, Biddulph (1970) found the mean birth interval to be 26 months. Unemployed fathers had children with shorter birth intervals than did employed fathers. Among the Wopkaimin, it is the reverse. The young fathers in the new villages are having their children close together and the unemployed in the old villages still have their traditional longer birth intervals. The Mt. Obree birth intervals show none of these differences as hardly anybody is 'employed', in the Western sense. It is a well known fact that shorter birth intervals are associated with low birth weight and high child morbidity and mortality (Wolfers *et al.*, 1975; Carlow and Vaidya, 1983). The Wopkaimin who are now experiencing decreasing infant mortality (Table 4.4) and shorter birth intervals at the same time since they have moved to the new villages find this difficult to accept.

Place of birth

The Wopkaimin started using the health institution for child birth in 1980 and minimized some of the obstetrical morbidity and mortality they faced using the traditional delivery huts. At the same time this exposed them to health education including family planning. Within four years they attained health institution births of 47% of all births by 1983, which is more than the national average of 42% health institution deliveries (Handbook on health statistics, Papua New Guinea, 1986). This has not increased further (Table 4.5) simply because the Company's health facilities are basically for its workers. By 1988 a joint Company and Government Hospital was opened and hopefully this will change the picture. The Mt. Obree women hardly ever use the health institutions for child birth simply

Table 4.5. *Percentage of health institution births among the Mt. Obree and Wopkaimin*

	Year								
Area	*1980*	*1981*	*1982*	*1983*	*1984*	*1985*	*1986*	*1987*	*1988*
Wopkaimin	15	10	33	47	35	35	28	38	31
Mt. Obree	2	0	0	3	2	0	4	1	2

The proportion of Wopkaimin health institution deliveries reached a peak in 1983 and has remained relatively the same. The Mt. Obree institution deliveries has remained low throughout.

because of communication problems. The very few that have health institution deliveries just happened to be in urban centres with their husbands when in labour. The village deliveries are associated with higher morbidity and mortality as reflected in their high infant mortality (Table 4.4). They would therefore wish to have more children to compensate for this loss.

Completed fertility rate

The average family size of Wopkaimin women who have completed their family in the traditional village is about four children. This is lower than the average Papua New Guinea figure of 5.36 (Shadlow, 1984). However, the Wopkaimin figure could be higher. The newborn babies are not given names for months and if they die before being named then they would not be regarded as live births. The Mt. Obree average family size is about 6 children. The average Papua New Guinea rural family size is larger than the urban family size (Shadlow, 1984). In the rural areas children are regarded as an asset to help with gardening, hunting and continuation of the tribe. In urban areas children are seen more as a liability to be fed, clothed and educated.

The Wopkaimin who have been suddenly exposed to a cash economy in 1982 (Table 4.6) show none of the urban family size trend as yet. The Mt. Obree people who are hardly exposed to the cash economy (Table 4.6) still maintain the rural larger family size.

Natural increase rate

After the negative natural increase rate (NIR) in 1983 due to the measles epidemic the NIR has been phenomenal for the Wopkaimin (Table 4.7).

Table 4.6. *Cash income to the two areas in Kina*[a] *and as per capita*

Year	*1982*	*1983*	*1984*	*1985*	*1986*	*1987*	*Total*
Wopkaimin	99 024.85	126 365.95	381 784.59	310 423.82	329 239.66	399 204.03	1 646 042.80
Per capita	142.28	174.78	463.89	365.20	358.65	412.21	1 654.31
Mt. Obree	4 000	9 000	15 000	4 000	15 000	—	47 000
Per capita	3.42	7.58	12.44	3.27	12.07	0	37.24

Source: Mr J. Ransley, Senior Liaison Officer, Provincial Affairs (Ransley, 1988) Department, PNG. (Wopkaimin).
– Mr Steven, DPI Extension Officer, Central Provincial Government (Batu, 1988), PNG, (Mt. Obree).

[a] Exchange rate K1 = US$1.1810.

For the Wopkaimin income consists of Special Mining Lease (SML) payments and lease for Mining Purpose (LMP). The royalty money only goes to Bultem & Finalbin Villages. All this is paid by the Mining Company to the villagers through the Government District Office in Tabubil. Income from the few who work for the company and sales of garden vegetables on Saturdays are not included here.

The Mt. Obree income is from the sale of chilli seeds to the Department of Primary Industry, Central Provincial Government (CPG). There were no funds for 1987 as the CPG was suspended. A very small income from few relatives who remit cash from town and sale of local betelnuts and vegetables is not included.

Table 4.7. *Natural increase rate (NIR)[a] in the two areas*

	Year						
Area	*1982*	*1983*	*1984*	*1985*	*1986*	*1987*	*1988*
Wopkaimin	2.9	−1.1	6.2	7.5	6.1	6	5.4
Mt. Obree		0.9	2.6	1.9	0.8	0.8	0.7

[a] A population increase or decrease in a given year as a percentage of the base population.
The Wopkaimin NIR increased substantially in 1984 and there has been a slight decrease since. The Mt. Obree NIR has remained relatively the same.

Table 4.8. *Family planning utilization rate[a] in the two areas*

	Year						
Area	*1982*	*1983*	*1984*	*1985*	*1986*	*1987*	*1988*
Wopkaimin	0	1.2	2.2	2.6	5.7	16.2	22.4
Mt. Obree		4.1	3.6	2.2	2.5	3.5	3.2

[a] Number of women using family planning method per 100 women of child bearing age (15–49 years).
The family planning utilization rate increased exponentially for the Wopkaimin. The Mt. Obree family planning rate has remained relatively the same.

This is the result of better socio-economic environment and available health services as reflected in the falling infant mortality rate (Table 4.6) but high crude birth rate (Table 4.1). Growth rate has not been included because of the difficulty of pinpointing the very mobile migrant population. The June 1988 census gave a phenomenal 141.2% growth rate for the Wopkaimin. The Mt. Obree people who experience no socio-economic changes have a low NIR (Table 4.7) and with very little population movement this could be taken as the growth rate also.

Family planning

Wopkaimin women started using Western family planning methods by 1983 (Table 4.8). Now more Wopkaimin women (22.4%) of child bearing age group use family planning than the average (8%) in Papua New Guinea. (Handbook on health statistics, Papua New Guinea, 1984). This could have contributed to the slight fall in CBR among the Wopkaimin.

The Mt. Obree utilization of family planning is minimal. With the very high IMR (Table 4.4) in both areas, family planning should be handled with great sensitivity. Experience in other countries has shown that, apart from broad socio-economic factors, people are not readily amenable to family planning unless they see that their children survive beyond infancy.

Other contributing factors to high fertility

The Wopkaimin are now a minority in their own villages (Table 4.9). This threatens them and consequently they wish to have as many children as possible to protect their land and share the riches that are now being mined in their land by outsiders. The current cash income from the land lease and mining royalty (Table 4.6, Figure 4.3) also contributes to this high fertility. The authorities, to be fair, distribute this equitably on a per capita basis. The villagers on the other hand see that if they have more family members they receive more money and thus strive to have more babies. It acts as an inventive for larger families. The Mt. Obree people do not face these changes in their invironment and therefore have no significant changes in their fertility rate.

Conclusion

Two Papua New Guinea communities have been studied. The Wopkaimin who have experienced major socio-economic and environmental changes show a markedly higher fertility rate. Economic development has resulted

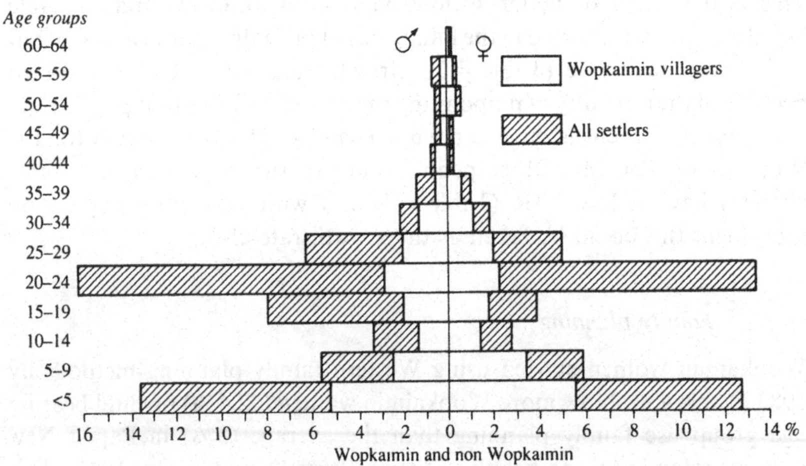

Figure 4.3. Total Wopkaimin villagers plus all settlers.

Table 4.9. *The demographic impact of migrants to Wopkaimin villages, June 1988*

Wopkaimin villages	*Wopkaimin*		*Atemkit*		*Kevorabip*		*Settlers*			*Wopkaimin villagers as percentage of total 'village' population*
	M	*F*	*M*	*F*	*M*	*F*	*M*	*F*	*Total*	
Wangbin	82	95	—	—	—	—	551	369	1097[a]	16.1%
Migalsimbip	88	84	—	—	—	—	86	73	331	52%
Bultem	114	121	38	29	6	6	105	71	490	48%
Finalbin	63	58	6	5	21	27	46	36	262	46%
Total	347	358	44	34	27	33	788	549	2180	32.3%

[a] A massive increase in the Wangbin population by 113% took place over the last six months.

Note Atemkits tend to live with Bultems and Kavorobips with Finalbins. Apart from Migalsimbip, the Wopkaimin are now a minority in their own villagers.

in the decline of traditional authority and men sleeping with their wives in the same house, which has resulted in significantly shorter birth intervals and a high general fertility rate. The better infrastructure and health services have resulted in a falling mortality rate but with minimal change to the crude birth rate despite the increased use of family planning. The threat posed by the large influx of migrants and the distribution of cash income from royalty and land lease, acts as an incentive for them to have high general fertility. The Mt. Obree people who have none of these socio-economic and environmental changes have experienced no change in their fertility rate.

Acknowledgements

We wish sincerely to thank Ok Tedi Mining Limited for their generous funding of the Wopkaimin part of the study, the University of Papua New Guinea for funding the Mt. Obree part of this work, and the Wopkaimin and Mt. Obree people, whose hospitality and assistance, despite many seemingly foolish questions, have enabled this work to continue.

References

Batu, S. (1988). *Department of Primary Industry Report*, Dorobisoro.

Biddulph, J. (1970). Longitudinal survey of children born in a peri-urban Papuan village. *Papua New Guinea Med. J.*, **13**, 23–7.

Carlow, R. W. & Vaidya, K. (1983). Birth intervals and the survival of children to age five – some data from Nepal. *J. Trop Pediatr.*, **29**, 31–4.

Handbook on health statistics (1986). Papua New Guinea, Department of Health, Port Moresby.

Jackson, R. (1982). *Ok Tedi: The Pot of Gold.* Port Moresby: University of Papua New Guinea.

National Statistical Office (1983). *Provincial Data System, Rural Community Register, Western Province.* Port Moresby: Government Printer.

Ransley, J. (1988). *Brief For Minerals and Energy.* Provincial Affairs Department, Tabubil.

Schlencker, Rev. (1895). *London Missionary Society District Report, Port Moresby 1895.* The London Missionary Archives, ALX-31, New Guinea Collections, Michael Somare Library, University of Papua New Guinea.

Shadlow, J. (1982). *National Population Census, Central Province.* National Statistical Office, Port Moresby.

Shadlow, J. (1984). *Preliminary Fertility Estimates derived from 1980 Census Data for Geographical Sub-Divisions of Papua New Guinea (working paper No. 7).* National Statistical Office, Wards Strip, Port Moresby.

Taufa, T., Lourie, J. A., Mea, V., Sinha, A. K., Cattani, J. & Anderson, W. (1986). Some obstetrical aspects of the rapidly changing Wokaimin society. *Papua New Guinea Med. J.*, **29**, 301–7.

United Nations World Population Chart, 1985. Population Division, Department of International Economic and Social Affairs, U.N. Geneva.

Wolfers, D. & Scrimshaw, S. (1975). Child survival and intervals between pregnancies in Guayaquil. *Ecuador. Popul Stud.*, **29**, 479–96.

5 *The cultural context of fertility transition in immigrant Mennonites*

J. C. STEVENSON AND P. M. EVERSON

Introduction

In the 1870s, Mennonites from the village of Alexanderwohl in the Ukraine Molotschna Colony immigrated to the United States and settled in choice farming areas around Goessel and Meridian, Kansas, and Henderson, Nebraska (McQuillan, 1978; Crawford & Rogers, 1982; Stevenson *et al.*, 1989; see Figure 5.1). They arrived with large families to a country already in the midst of fertility decline (Vinovskis, 1981), and within one generation they also were reducing family size (Stevenson *et al.*, 1989). The recency of this fertility transition provides an opportunity to assess the relative importance of education, occupational opportunities outside the home, and the labour value of children in stimulating the reproductive decline for the Kansas and Nebraska Mennonites. This analysis is based on Caldwell's (1982) wealth flows model as modified by Handwerker (1986a, b, c).

Foreign colonists who had settled in Russia since 1763 were exempt from military and civil service 'in perpetuity' (Rempel, 1974). However, reforms in the 1860s resulted in the cancellation of this exemption and institution of compulsory military service. The most politically conservative portion of the Mennonite population (eventually one-third) began to emigrate to the United States and Canada in the 1870s. Many selected choice farmland in Kansas and Nebraska with the aid of railroad agents, and by the 1880s Mennonites were some of the most successful farmers, experiencing relatively few foreclosures during the severe droughts of the 1890s (McQuillan, 1978).

The Mennonite group which immigrated was also the most religiously conservative of the Molotschna Colony (Rempel, 1974). Pollack (1978; also discussed in Hurd, 1985) compared Old Amish, Mennonites and non-Mennonites and noted the existence of a positive association between religious conservatism and the number of live births per completed family in Plain City, Ohio. The most conservative Amish still proscribe birth control, and to date, fertility decline is not evident for the Amish (reviewed in Stevenson *et al.*, 1989). One can infer, at least initially, that the earliest

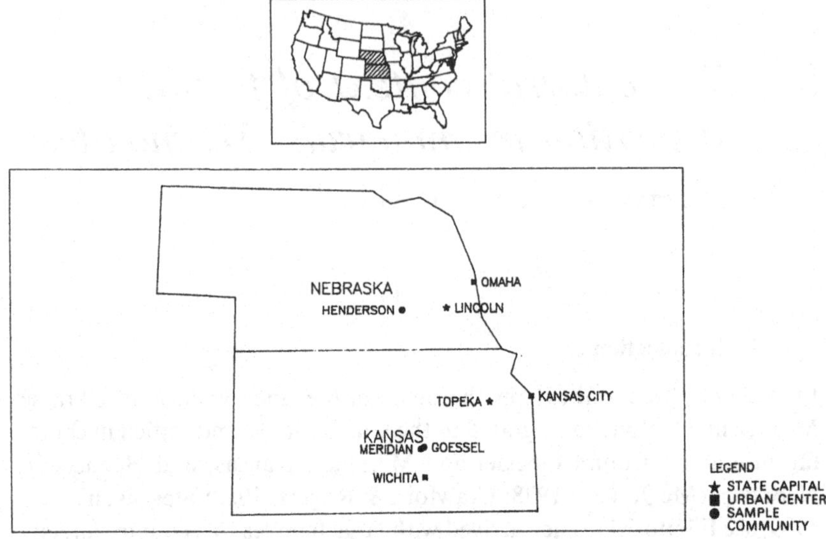

Figure 5.1. Location of Mennonite sample communities.

Mennonite immigrants had similar attitudes about optimum family size. A conservative estimate of mean completed family size for the immigrant Mennonites was obtained by examining the limited demographic data for the Mennonite families who travelled by ship and settled in Kansas during the years 1872–1875 (Haury, 1986). There were 230 families which included mothers aged 40 to 49 years. These women had completed or almost completed their families, and most older children would still be part of the migrating household and on the family lists. The number of children per family ranged from zero to twelve, and the average number of children per family was 5.2 ($\pm\sigma$, $\sigma=2.5$). Although comparable data for Nebraska are not readily available, seventeen families with mothers aged 40–49 years migrated to the Nebraska community of Henderson during 1874–1875 (Voth, 1975). Family size ranged from three to twelve children, and mean completed family size was 6.6 ($\pm\sigma$, $\sigma=2.4$) children. Therefore, family sizes were large for the immigrant Mennonites, averaging at least five to six children. In addition, Coale *et al.* (1979) note that marital fertility in those areas surrounding the Molotschna Colony was extremely high, and fertility decline did not occur until after 1900.

Stevenson *et al.* (1989) have also documented mean completed family sizes for married women from Goessel and Meridian, Kansas, and Henderson, Nebraska, born during the decades from 1870 to 1939. A decline in fertility is apparent in all three communities, decreasing from a

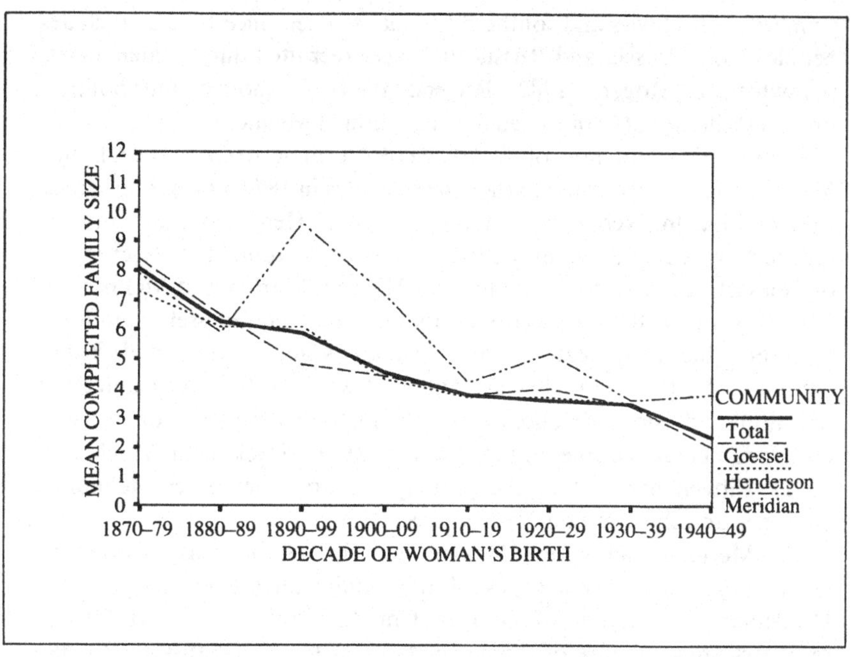

Figure 5.2. Mennonite mean completed family size. *Source*: Stevenson *et al.* (1989).

high of seven to eight children among women born in the decade 1870–1879, to three to four children among women born in the decade 1930–1939 (see Figure 5.2). The decline in mean completed family size from seven to eight children to three to four children occurred for those women born during the late 1800s; mean completed family size stabilized between three to five children for women born from 1900 to 1939. Overall, mean completed family size decreased by 53.1% between the 1870 cohort and the 1910 cohort. Of this decline, 41.9% occurred for women born during the decade 1880–89, and 9.09%, 31.8% and 17.2%, for women born during the decades 1890–99, 1900–09, and 1910–19, respectively. There has been only a 4.0% decline in mean completed family size for women born in the decades following (1920–29 and 1930–39). Thus, fertility transition began with women born during the period 1880–1919. The objective here is to explore how the cultural milieu in the New World, during this period, led to a change in behaviour resulting in smaller family sizes.

Populations studied

The individuals residing in Goessel and Meridian, Kansas, and Henderson, Nebraska, are the descendants of sixteenth and seventeenth century Dutch primarily, plus Swiss and southern Germans. Their ancestry also includes peoples from Prussia and Russia that were recruited during later travels (Crawford & Rogers, 1982; Rogers, 1984). Economic and political upheavals in the 1860s preceded the migration of Mennonites out of Russia (Rempel, 1974), and most of the village of Alexanderwohl, of the Ukraine Molotschna Colony, came to the United States in 1874. One group settled west of Lincoln, Nebraska, in today's town of Henderson, and another group, New Alexanderwohl, settled in rural areas around the present town of Goessel, Kansas, in the counties of Harvey, Marion and McPherson.

In this study, 83% of the participants from the Goessel community represent the congregations of Alexanderwohl, Goessel and Tabor churches. All three churches are affiliated with the General Conference Mennonite division, a moderate group which only requires adherence to the fundamental doctrines of the Mennonite faith (Dyck, 1981; Van Meter, 1972). Fifteen per cent of the participants are from other Mennonite churches and 2% are non-Mennonites.

The Meridian community, located near Hesston, Kansas, includes the most conservative Mennonites of this study; all are members of the Holdeman, or Church of God in Christ. Ethnically, the Holdeman members are a mixture of the descendants of the 1870 Prussian-Russian immigrants and American Mennonites from Indiana and Pennsylvania (Crawford & Rogers, 1982).

The Henderson community consists of descendants from the Molotschna colony villages, including Alexanderwohl (Crawford & Rogers, 1982; Rogers, 1984; Voth, 1975). Individuals represent three Henderson congregations: the Bethesda Church (74%) affiliated with the General Conference Mennonite division, the Mennonite Brethren (16%), and the Evangelical Mennonite Brethren (3%). Seven per cent are in a miscellaneous category which includes both non-Mennonites and out-of-state Mennonites.

Data and methods

Information on education, occupation and family size for individuals from the communities of Goessel and Meridian, Kansas, and Henderson, Nebraska, was obtained as part of a larger interdisciplinary ageing study described in Crawford & Rogers (1982). Interview and questionnaire data were collected at health clinics in the Goessel and Meridian churches in Kansas, during January 1980, and in the Bethesda Church of Nebraska during January 1981. Although numerically a small sample, the Meridian group included 54% of the adults residing in the community and 100% of the families (Sirijaraya, 1984). Eighty-three per cent of the participants from the Goessel community represent the Alexanderwohl, Goessel and

Tabor churches; this sample includes 47% of the entire Alexanderwohl church membership. The town of Henderson consisted of 971 persons, and 547 individuals (56%) participated in this study.

The similarity in rural setting, ethnicity, lifestyle, and virtual identity in the graphs depicting change in completed family size over time for the three communities (Stevenson *et al.*, 1989) are the justification for pooling the occupational and educational data for individuals from the three communities. Mennonite cohorts born during the decades 1900–1940 were compared by decade for relative proportions of: persons who completed high school, and the dependency of males and females on farming versus non-farm occupations.

Historical accounts were used to reconstruct educational and occupational opportunities for Mennonites in Russia; census data and historical accounts were used to reconstruct the economic climate and educational and occupational opportunities for Kansas and Nebraska Mennonites of the late 1800s and early 1900s.

Discussion

Transition theory

Central to contemporary demography is research on the modern demographic transition (Caldwell, 1976, 1982; Cleland & Wilson, 1987). The theoretical basis of this work as originally formulated by Notestein (1945, 1953) states that traditional societies exhibit high fertility because of 'high mortality, the lack of opportunities for individual advancement, and the economic value of children' (Caldwell, 1976: p. 324). With modernization and urbanization, attitudes change, and fertility decreases.

However, factors associated with urbanization and modernization do not consistently correlate with a reduction in fertility (Handwerker, 1986b, c). Nor does a mortality decrease necessarily precipitate a fertility decline (Taylor *et al.*, 1976; Van de Walle, 1986), and education level is usually but not always correlated with fertility levels (Cochrane, 1979; Graff, 1979). Thus, how, and why, a specific society shifts from a high to a low fertility regime is not yet predictable nor well-understood.

Caldwell (1976, 1978, 1981, 1982) has combined sociological and economic approaches in his 'wealth flows' theory. He argues that in pre-transitional societies with familial production, net goods and services flow from children to parents and grandparents. When this flow of wealth reverses direction (parents to children), which is typical for capitalist modes of production, then the economic value of children declines to zero and attitudes eventually favour smaller family size. Caldwell argues 'that the primary determinant of the timing of the onset of the fertility transition is the effect of mass education on the family economy' (1982: p. 301). In particular, education restructures family relationships by reducing the child's potential for work within the family, increasing the child's costs, indoctrinating for Western middle class values and improving the child's status within the family. New attitudes follow.

Many studies are providing support for general wealth flows theory (e.g., Handwerker, 1986b), but Handwerker (1986c) disagrees with some of the particulars of Caldwell's theory, specifically, the hypothesis that mass education is the determinant of the timing of fertility decline (Caldwell, 1980, 1982). Handwerker (1986c: p. 401) argues that Caldwell's emphasis on education fails to explain the reason why people change their way of thinking and 'rests on the covert tautology that attitude change determines behavioral change and, hence, can never explain why values change'. (See also Handwerker, 1986a, b.) Thus, Handwerker extends Caldwell's model by suggesting that fertility decline does not result solely from the onset of mass education but is a response to the 'conjunction of mass education with changes in opportunity structure that increasingly reward educationally acquired skills and perspectives' (Handwerker, 1986c: p. 402). In other words, fertility transition is most likely to follow when there are increasing non-agricultural job opportunities which reward formal schooling and skill training.

Handwerker (1986c) also expressed the differences between the two theories in terms of formal models, quantified all of his variables, and then applied his ideas in a regression analysis of data from 86 countries for 1960–1980. He found that neither education (as measured either by the percentage of the population literate, ca. 1980, or by the primary and secondary school enrollment as a percentage of age group, ca. 1960) nor mortality transition (as measured by infant mortality, ca. 1980, or 1960, and favoured by many demographers (e.g., Preston, 1978) as the trigger for fertility decline) accounted for fertility transition.

The focus of interest here is on the fertility transition which occurred after the Mennonites settled in the United States Midwest in the early 1870s (Stevenson *et al.*, 1989). The variables cannot be measured as precisely for historically derived data; however, the Caldwell model as modified by Handwerker can provide a general framework for analysis. Figure 2 provides a summary of change in mean completed family size for women from the three communities of Goessel and Meridian, Kansas, and Henderson, Nebraska. The initial decrease occurs for women born in the decade 1880–89, and continues with those born in the decades 1890–99, 1900–09, and 1910–19, after which time the decrease continues but at a slower rate. Women born during these decades would be bearing most of their children from the early 1900s until the Great Depression. Thus, it is necessary to examine the economic, occupational and educational climate prior to the decline, both in Russia during the 1850s and 1860s, and in Kansas–Nebraska during the 1870s, 1880s and 1890s, and then compare that to the period from 1890 to 1939 when family size declines.

A consideration of mortality transition is excluded from this study. Lin & Crawford (1983) recently analysed mortality data from the Alexanderwohl Mennonites with the use of a time series model for the period 1890 to 1972. Their analysis considered raw mortality figures, unadjusted for age or sex. Throughout this period the raw number of mortality events has remained nearly the same. Therefore, if mortality transition for the Alexanderwohl Mennonites did take place, it was prior to 1890. (For a review of the complexity of interaction between components of mortality and fertility during the European demographic transition see Van de Walle, 1986.)

Education

The Mennonite colonies in Russia were founded with schools, but the Russian government took no interest in the Mennonite schools until the late 1870s (Rempel, 1974: pp. 40–1). The Mennonites had already made major improvements in 1843. The administration of schools was moved from the Mennonite clergy and elders and placed under the supervision of the Agriculture Unions. Reforms included substituting High German for Low German (the dialect), testing teachers, improving school buildings, and lastly, requiring school attendance for boys and girls between the ages of six and fourteen.

Thus, when Mennonites moved to the Midwest they established both churches and schools (Smith, 1981). The Bible and German language remained important components of Mennonite education. 'Soon a belt of elementary and secondary schools operated by the Mennonites flourished in all Mennonite communities between Newton, Kansas, and Winnipeg, Manitoba' (Smith, 1981: p. 433). Secondary schools based on the pattern of the Russian secondary schools (Zentralschule) led to the founding of Halstead Seminary in 1882 which was succeeded by Bethel College in 1887. Teacher preparation was also promoted, and schools were built in all the larger Mennonite communities. However, public schools replaced parochial schools gradually.

Three schools were started at the New Alexanderwohl community in the winter of 1874–1875 (Van Meter, 1972). Later, students began to attend public schools, because laws were still lenient and Bible instruction was permitted. In addition, public school terms were only three to five months

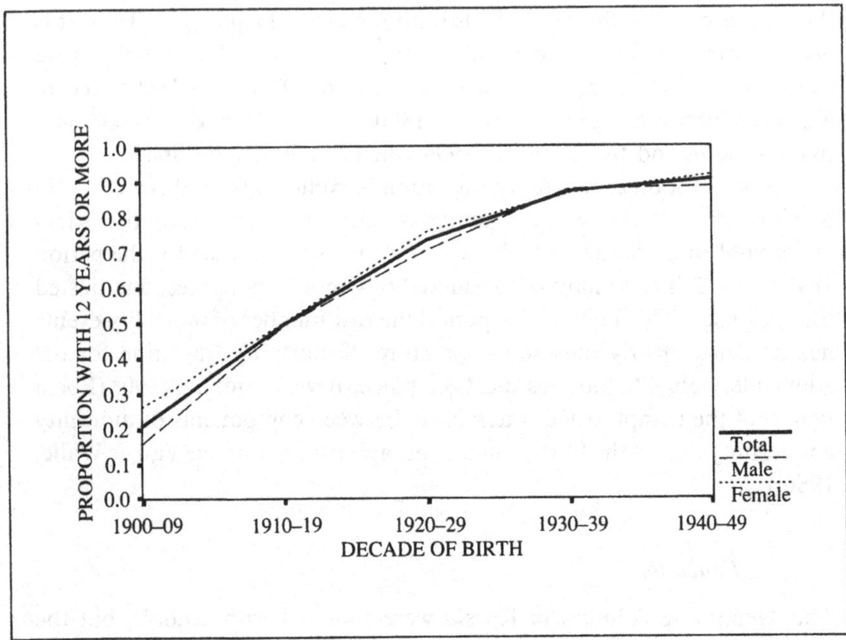

Figure 5.3. Mennonite educational attainment.

long so that church schools could operate during the vacation months. However, Kansas state law changed in 1911 when school terms were lengthened to seven months, and teaching German was prohibited after World War I. Thus, most Mennonite children were in the public schools by the 1920s.

Education *per se* does not seem to be the catalyst for fertility decline in the Mennonites. Education for both sexes preceded the decline, and the acculturation to middle class values which may have occurred as a result of attendance at American public schools took place after the decline was underway. However, in Figure 5.3 it is apparent that attendance at American public schools increased dramatically for both sexes in the early part of this century. Thus, the fertility decline already underway was likely reinforced by increasing exposure to American middle class values.

Economic climate and occupational structure

In 1841, out of 2733 families of the Molotschna colony, only 1033 (38%) were land-owning farmers, and the remaining 1700 (62%) families were small tenant farmers or involved in other businesses or trades (Rempel,

1974: 24–25). By 1860, almost two-thirds of those in 57 communities of the Molotschna colony were landless cottagers with or without ownership of their homes. Tradesmen and craftsmen were becoming too numerous and were unable to migrate easily to other areas of Russia, because they were also partially responsible for the land tax shared communally by the entire colony. Thus, requests were made to redistribute the lands of the colony in a more equitable fashion.

Additionally, a religious reform movement spread to the Mennonites and resulted in the formation of the Brethren Church (Rempel, 1974). The new church received the most support from the landless and poorer Mennonites, and the Russian government gave the church full recognition and directed that there should be a solution to the land disputes. Finally, in 1866 the land was redivided, the landless were enfranchised, and a communal fund was created for appropriating more land. Each settlement was expected to provide for future generations. Thus, a mechanism for acquiring more land was installed. However, new land was difficult to acquire (Duerksen, in Rogers, 1984), and this pressure, added to the removal of the military service exemption and Russianization of the schools in the 1870s and 1880s, forced many families to migrate to North America.

Immigrants were encouraged to settle in Kansas, and on March 9, 1874, the Kansas legislature passed an act exempting the Mennonites from military service (Zornow, 1957). A brutal winter during 1874–1875 coupled with a heavy infestation of grasshoppers during the summer of 1874 discouraged many of the resident farmers. The immigrant Mennonites were not daunted, because the presence of grasshoppers implied good soil, and they were already familiar with dryland farming (Davis, 1976). The Mennonite Bernard Warkentin, brought the seed of Turkey Red winter wheat to Kansas in 1874 (the only wheat harvested before the grasshopper infestation) (Zornow, 1957; Smith, 1981). Plant breeders have since developed all Kansas wheat strains from this seed. Mennonite tillage methods were also rapidly adopted.

Thus, Mennonite immigrants selected some of the best land in Kansas and Nebraska and were soon successful farmers (McQuillan, 1978). Initially, fields were allocated to everyone on the Russian pattern of a centralized set of homesteads (village settlement) roughly equidistant from individual family fields (Quadagno & Janzen, n.d.). By 1882, lands were redistributed so that all could build homesteads on their field lands which is the American custom. Thus, the village settlement eventually disappeared, although individuals now had much more personal freedom to take risks, buying and selling property. A disadvantage was that families were now responsible for providing for their offspring themselves rather than having

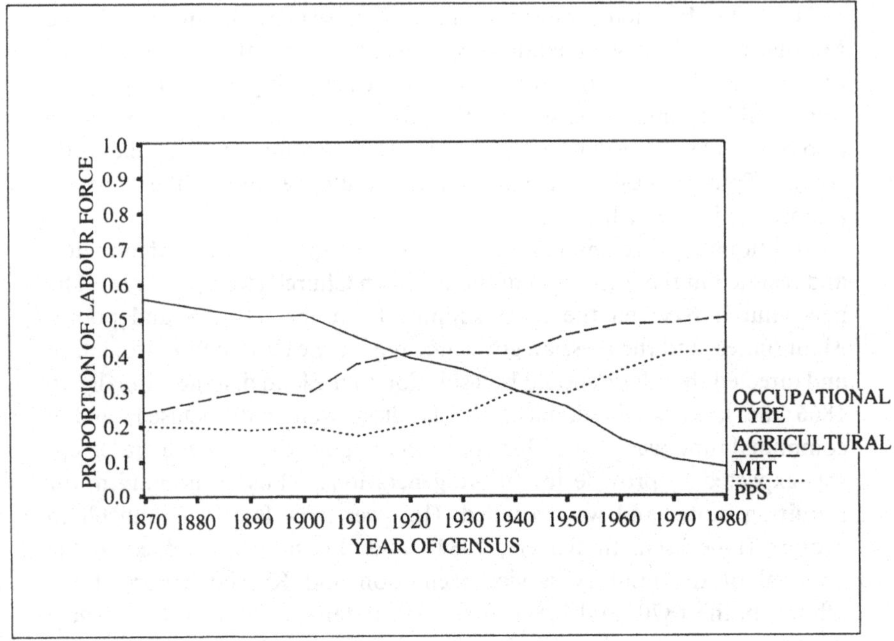

Figure 5.4. Kansas–Nebraska occupational structure, 1870–1980. *Source*: U.S. Bureau of Census, 1870–1980. Key: MTT – Manufacture/Trade/Transport. PPS – Personal/Professional Service.

access to the pooled resources of the colony to buy new lands for expansion. In addition, Mennonite kin groups considered land and estate capital to be the property of the entire group and that resources were to be shared equally among the heirs (Quadagno & Janzen, n.d.).

By the 1890s all the land was taken, and profits from land speculation were gone (McQuillan, 1978). From 1893–1896 there was a drought resulting in crop failures, although Mennonite farmers were more successful than Swedish or French Canadian immigrant farmers in avoiding foreclosure. They were also investing in new farming technology and emphasizing wheat farming and dairying. Thus, from 1895 to 1905 many Mennonites were expanding the size of their farms, and from 1905 to 1915 harvests were above average. Farm price increases after World War I began in 1914 and did not decline again until after 1925, but a post-war depression began in 1919, and crop yields varied in the years 1915–1925.

The proportion of the labour force in agriculture is considered an indirect measure of industrialization and occupational opportunities outside of farming (Handwerker, 1986c). Figure 5.4 presents the proportions of the Kansas–Nebraska labour force in: (*a*) agriculture, (*b*) manu-

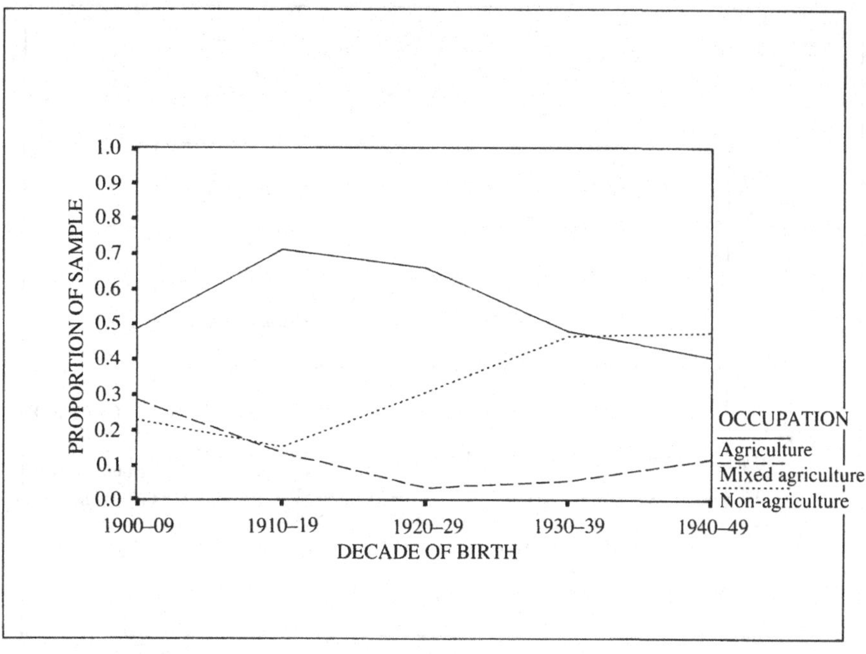

Figure 5.5. Mennonite occupational structure – male.

facturing, trade and transportation, and (*c*) personal and professional service. The proportion of the labour force engaged in agriculture declined steadily from 1900 to 1980, and opportunities in manufacturing, trade and transportation, and in the personal and professional services, increased in the same manner. The increasing number of occupational opportunities outside of agriculture during the time of fertility decline coincided, generally, with the change from labour intensive, to capital intensive, agriculture, as noted above.

An increasing proportion of Mennonite men and women born after 1900, particularly those born during the decades 1910–19 and 1920–29 took non-agricultural occupations (see Figures 5.5 and 5.6). Individuals born during those decades would be entering the occupational structure during the Depression and World War II.

Handwerker (1986c: p. 400) suggests that 'fertility transition in the contemporary world comes about when personal well-being is determined less by personal relationships than by formal education and skill training'. He argues that this transformation is triggered when occupational opportunities in the external labour market require increased education and skills. The specifics of his argument are not supported here. Non-

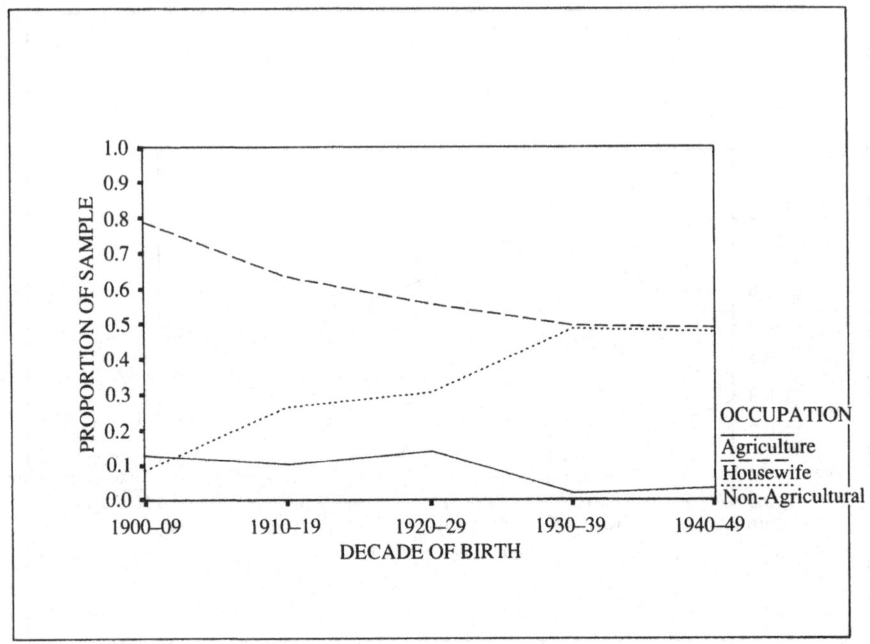

Figure 5.6. Mennonite occupational structure – female.

Mennonite Kansas and Nebraska natives began exploiting increasing occupational opportunities outside of agriculture after 1900. The proportion of the labour force in agriculture dropped steadily after 1900, whereas most Mennonites, men and women, remained in agriculture, or occupations based on 'personal relationships', i.e., housewives. Increases in the proportions of Mennonites entering non-agricultural occupations occurred first among women as they entered the non-agricultural labour force in the 1930s. The number of men entering the non-agricultural labour force did not increase until later, in the 1940s. Both of these dates come at the tail end of the period of fertility decline.

Conclusions

Completed family size for women born during the 1870s averaged seven to eight children. These women were bearing children during a very prosperous time for Mennonite farmers. Farms expanded in size, crop yields were above average, and new technology was added.

Mean completed family size for women born during the 1880s decreased 22.2% marking the beginning of the fertility transition. Although families were started during economically secure times, World War I occurred during the later years of childbearing and was followed by an economic depression. This was also the first generation born after the switch to American-style homesteads. Thus the shift to a low fertility regime coincided with changes in social relations, and with changes in the regional economy. The switch from a communally shared system of land tenure to the American system of individual family farms brought about important changes in the relations of production. Individual families no longer had the pooled resources of the community to rely upon, and the economic security of offspring became a familial rather than a communal responsibility. Exacerbating this situation after World War I was the increasing difficulty in obtaining new farm land. Thus, the expense of providing for offspring increased dramatically for parents born after 1880.

This study provides general support for Caldwell's wealth flows model. The initiation of the fertility decline follows an increase in the 'cost' of children. Handwerker is also partially supported because changing social relations accompany the fertility transition. However, neither mass education nor external occupational opportunities requiring special skills appear to trigger the decline, although both surely reinforced the reduction in fertility. Thus, this case study of a transplanted Russian Mennonite immigrant population illustrates the complexity and multicausal nature of fertility transition, and it is only through more study of specific instances of fertility decline that a base can be built from which to refine the models and ultimately understand the nature of demographic transition.

Acknowledgements

This research was supported in part by N.I.H. grant AGO1646-03.

We are grateful to the Goessel, Meridian and Henderson congregations, without whose support and participation this study would not have been possible.

We wish to thank Dr. M. H. Crawford for generously providing access to the data, Dr. L. Rogers and Ms. M. Uttley for their assistance in data collection and analysis, and Mr. L. Tedrow of the Demographic Research Lab, Western Washington University, for his assistance in preparing the figures.

References

Caldwell, J. C. (1976). Toward a restatement of demographic transition theory. *Population and Development Review*, **2**, 321–66.

Caldwell, J. C. (1978). A theory of fertility: from high plateau to destabilization. *Population and Development Review*, **4**, 553–77.

Caldwell, J. C. (1980). Mass education as a determinant of the timing of fertility decline. *Population and Development Review*, **6**, 225–55.

Caldwell, J. C. (1981). The mechanisms of demographic change in historical perspective. *Population Studies*, **35**, 5–27.

Caldwell, J. C. (1982). *Theory of Fertility Decline*. New York: Academic Press.

Cleland, J. & Wilson, C. (1987). Demand theories of the fertility transition; an iconoclastic view. *Population Studies*, **41**, 5–30.

Coale, A. J., Anderson, B. A & Harm, E. (1979). *Human Fertility in Russia Since the Nineteenth Century*. Princeton, New Jersey: Princeton University Press.

Cochrane, S. (1979). Fertility and education: what do we really know? *World Bank Staff Occasional Papers, No. 26*. Baltimore, MD.: Johns Hopkins University Press.

Crawford, M. H. & Rogers, L. (1982). Population genetic models in the study of aging and longevity in a Mennonite community. *Social Science and Medicine*, **16**, 149–53.

Davis, K. S. (1976). *Kansas: A Bicentennial History*. New York: W. W. Norton and Company.

Dyck, C. J. (1981). *Mennonite History: A Popular History of the Anabaptists and Mennonites*. Scottdale, Pennsylvania: Herald Press.

Graff, H. J. (1979). Literacy, education, and fertility, past and present: a critical review. *Population and Development Review*, **5**, 105–40.

Handwerker, W. P. (1986a). Culture and reproduction: exploring micro/macro linkages. In *Culture and Reproduction: An Anthropological Critique of Demographic Transition Theory*, ed. W. P. Handwerker, pp. 1–28. Boulder, Colorado: Westview Press.

Handwerker, W. P., (ed) (1986b). *Culture and Reproduction: An Anthropological Critique of Demographic Transition Theory*. Boulder, Colorado: Westview Press.

Handwerker, W. P. (1986c). The modern demographic transition: an analysis of subsistence choices and reproductive consequences. *American Anthropologist*, **88**, 400–17.

Haury, D. A., (ed.) (1986). *Index to Mennonite Immigrants on United States Passenger lists, 1872–1904*. Newton, Kansas: Mennonite Library & Archives.

Hurd, J. P. (1985). Sex differences in mate choice among the 'Nebraska' Amish of central Pennsylvania. *Ethology and Sociobiology*, **6**, 49–57.

Lin, P. M. & Crawford, M. H. (1983). A comparison of mortality patterns in human populations residing under diverse ecological conditions: a time series analysis. *Human Biology*, **55**, 35–62.

McQuillan, D. A. (1978). Farm size and work ethic: measuring the success of immigrant farmers on the American grasslands, 1875–1925. *Journal of Historical Geography*, **4**, 57–76.

Notestein, F. W. (1945). Population: the long view. In *Food for the World*, ed. T. W. Schultz, pp. 36–57. Chicago, Illinois: University of Chicago Press.

Notestein, F. W. (1953). Economic problems of population change. In *8th International Conference of Agricultural Economists, 1953*, pp. 15–18. London: Oxford University Press.

Pollack, R. (1978). Genetic demography of an Amish–Mennonite population. *American Journal of Physical Anthropology*, **48**, 427.

Preston, S. H., (ed.) (1978). *The Effects of Infant and Child Mortality on Fertility*. New York: Academic Press.

Quadagno, J. & Janzen, J. M. (n.d.) Old age security and the family life course: a case study of nineteenth century Mennonite immigrants to Kansas. Unpublished Manuscript, Department of Anthropology, University of Kansas, Lawrence, KS.

Rempel, D. G. (1974). The Mennonite Commonwealth in Russia, a sketch of its founding and endurance, 1789–1919. *The Mennonite Quarterly Review*, **48** 5–54.

Rogers, L. A. (1984). Phylogenetic identification of a religious isolate and the measurement of inbreeding. Unpublished Ph.D. Dissertation, University of Kansas, Lawrence, Kansas.

Sirijaraya, S. (1984). Inbreeding in the Meridian Mennonites of Kansas. Unpublished M.A. Thesis, University of Kansas, Lawrence, Kansas.

Smith, C. H. (1981). *Story of the Mennonites*, 5th edn., revised by C. Krahn. Newton, Kansas: Faith and Life Press.

Stevenson, J. C., Everson, P. M. & Crawford, M. H. (1989). Changes in completed family size and reproductive span in Anabaptist populations. *Human Biology*, **61**, 99–115.

Taylor, C. E., Newman, J. S. & Kelly, N. U. (1976). The child survival hypothesis. *Population Studies*, **30**, 263–77.

Van de Walle, F. (1986). Infant mortality and the European demographic transition. In *The Decline of Fertility in Europe*, ed. A. J. Coale & S. C. Watkins, pp. 201–33. Princeton, New Jersey: Princeton University Press.

Van Meter, S. (1972). *Marion County, Kansas: Past and Present*. Hillsboro, Kansas: M.B. Publishing House.

Vinovskis, M. A. (1981). The fertility decline in the West as a model for developing countries today: the case of nineteenth century America. In *Fertility Decline in the Less Developed Countries*, ed. N. Eberstadt, pp. 228–53. New York: Praeger Publishers.

Voth, S. E., (ed.) (1975). *Henderson Mennonites, From Holland to Henderson*. Henderson, Nebraska: Henderson Centennial Committee, Service Press, Inc.

Zornow, W. F. (1957). *Kansas, a History of the Jayhawk State*. Norman, Oklahoma: University of Oklahoma Press.

6 *Inter-relationships between consanguinity, religion and fertility in Karnataka, South India*

A. H. BITTLES, A. RADHA RAMA DEVI AND N. APPAJI RAO

Introduction

When considering factors which act as major influences on fertility, religion and consanguinity may feature prominently, especially among the populations of less developed nations. Unfortunately, in many such studies it has been the practice either to assess the effects of religion and consanguinity separately, or to ignore consanguinity altogether. For the investigation of possible inter-relationships between the three parameters South India provides an excellent study centre. In the four southern states, Andhra Pradesh, Karnataka, Kerala and Tamil Nadu, with a combined population in the 1981 Census of India of 164.1 millions, consanguineous marriages are strongly favoured and three major religions, Hinduism, Islam and Christianity, are practised. Therefore detailed studies into the relative roles and effects of religion and consanguinity on fertility are possible at local, state and regional levels.

The study population

Data have been collected on fertility and a variety of fertility-associated parameters in the cities of Bangalore and Mysore, the present and former capital of the state of Karnataka, as part of a prospective neonatal screening programme for the detection of inherited amino acid disorders (Appaji Rao *et al.*, 1988). Within three to five days of the birth of a child, details are obtained by interview from the mother on her age, the degree of genetic relatedness between husband and wife, their religion(s), and the number of liveborn and living children in the family. The consanguinity classes noted are: nonconsanguineous (with a coefficient of inbreeding in the offspring F=O), beyond second cousin ($F<0.0156$), second cousin ($F=0.0156$), first cousin ($F=0.0625$) and uncle–niece ($F=0.125$). All interviews are conducted by trained local staff in the mother's native language, usually Kannada, although owing to large-scale immigration into the two cities as a result of their rapid industrialization and expansion, Telugu, Tamil and Malayalam also are understood and spoken by many of those questioned.

The results of the present investigation are based on 99323 pregnancies delivered in hospitals located throughout the two cities between 1981 and 1988. Since an estimated 80% of all births in Bangalore and Mysore are hospital deliveries, and there is no preselection of subjects, the families sampled can be regarded as representative of the community as a whole. Almost inevitably there may have been under-sampling of women from itinerant or semi-itinerant households. Mothers from the highest socio-economic group, whose confinements mostly are in private clinics, also could be under-represented. Duplication may have occurred in the collection of family records because of women reporting with more than one pregnancy during the sampling period. However, specific questioning on this subject suggested a probable upper bias limit of between one and two per cent.

Patterns of religion and consanguinity

The religious profile of the study population is summarized in Table 6.1. As is typical in all four South Indian states (Sanghvi, 1966; Kumar *et al.*, 1967; Rao & Inbaraj, 1977; Radha Rama Devi *et al.*, 1981), Hindus form the majority community and account for 80.4% of the total with lower percentages of Muslims (15.9%) and Christians (3.6%). The minority Christians, mainly Roman Catholics and members of the Church of South India (Protestants), are believed to be relatively recent converts from Hinduism.

Details of the consanguinity classes in each of the three major religions are shown in Table 6.2. Although the women interviewed were city dwellers, it is probable that in a sizeable proportion of cases they and their spouses had migrated from the villages where their marriages were contracted. Owing to generally low literacy levels in Karnataka, with estimated female literacy of 19.8% in rural areas (Basu, 1987) and a consequent lack of written records, the data presented refer only to the present generation. As such, the levels of consanguinity cited must be regarded as minimal estimates of inbreeding in the population. In overall terms consanguineous marriages accounted for 34.4% of the total, with the highest rate of inbreeding among Hindus (36.5%) and lesser rates in Muslims (26.3%) and Christians (21.8%). The observed pattern of religious differentials in total consanguinity is similar to that found in

Table 6.1. *Religious profile of marriages studied*

	Number of marriages	*Proportion of marriages*
Hindu	79 883	0.8043
Muslim	15 790	0.1590
Christian	3 545	0.0357
Others	13	
Unspecified	92	
Total numbers	99 323	

Table 6.2. *Proportion of births by religion and consanguinity*

Marital type	*Coefficient of inbreeding (F)*	*Hindu*	*Muslim*	*Christian*
Nonconsanguineous	0	0.6014	0.7043	0.7608
Beyond second cousin	<0.0156	0.0449	0.0339	0.0351
Second cousin	0.0156	0.0159	0.0231	0.0126
First cousin	0.0625	0.1057	0.1707	0.0658
Uncle–niece	0.125	0.1989	0.0355	0.1048
Unspecified	—	0.0333	0.0325	0.0208
Total numbers		80 437	15 936	3 558

Andhra Pradesh and Tamil Nadu (Sanghvi, 1966; Rao, 1983) but in a small-scale study in Kerala, Muslims had higher levels of inbreeding than Hindus (Ali, 1968). In North India, although consanguinity is practised by Muslims (Basu, 1975), it is rigorously prohibited in Hindu marriages (Kapadia, 1958).

The most popular form of consanguineous marriage varied between the three religions. Hindus particularly favoured uncle–niece unions (19.9%) whereas Muslims preferentially contracted first cousin marriages (17.1%). Christians also opted for uncle–niece marriages (10.5%), probably an indication of the residual, pre-conversion influence of Dravidian Hindu marriage customs. While the coefficient of inbreeding value for the general population was $F=0.0286$, the equivalent levels for each religion were Hindus ($F=0.0317$), Muslims ($F=0.0155$) and Christians ($F=0.0174$), primarily reflecting the varying proportions of uncle–niece and first cousin marriages in each community. The patterns of first cousin marriage contracted by Hindus and Muslims also differ. Hindus almost exclusively favour cross-cousin marriages, predominantly mother's brother's daughter (MBD) (Hann, 1985a), whereas parallel as well as cross-cousin cousin marriages are common in the Muslim community.

In view of the current large proportion and overall number of Hindu uncle–niece marriages in these two urban Karnatakan populations, it is pertinent to note that, under the terms of the Hindu Marriage Act 1955, uncle–niece marriages are illegal (Kapadia, 1958). In Islamic law, although first cousin marriages ($F=0.0625$) and occasionally double first cousin ($F=0.125$) are favoured, uncle–niece marriages ($F=0.125$) are not permitted, yet such unions accounted for 3.6% of the observed Muslim total in Bangalore and Mysore. With regard to the Christian uncle–niece and first cousin marriages, if contracted between Roman Catholics Diocesan dispensation should have been a pre-requisite, but it is dubious whether this had been sought in all cases.

Maternal age profiles

In assessing the fertility of a population, maternal age, and more specifically the age at which first pregnancies are undertaken, is of considerable importance since it largely will determine the length of the female reproductive span. Maternal ages at delivery varied considerably, as shown in Figure 6.1 for all liveborn pregnancies ($n=99\,323$) and for first liveborn pregnancies only ($n=35\,173$). Most births were to mothers in the 20 to 24 years age group (42.7%), and 98.7% of all births were to mothers aged less than 40 years. The youngest mothers, three in number, were 12 years of age at the birth of their children. One mother claimed to have given birth at age 57 years but in Karnataka, where many persons do not know their age with absolute precision, this can most appropriately be regarded as a non-sampling error.

Prior to Independence the Child Marriage Act 1929 sought to prohibit girls marrying before the age of 14 years, although this ruling did not apply in the princely states such as Mysore, the forerunner of modern Karnataka. The Hindu Marriage Law 1955 prescribed 15 years as the minimum age for female marriage and subsequently this was raised to 18 years by the Child Marriage Restraint Act of 1978 (Kulkarni *et al.*, 1986). On the basis of the evidence presented in Figures 6.1 to 6.3 limited regard appears to have been paid to the legal conditions attached to minimum age at marriage. However since the 1950s female age at marriage in Karnataka gradually has risen (Caldwell *et al.*, 1983; Kulkarni *et al.*, 1986). In turn, this trend would be expected to reduce the reproductive span thus limiting fertility. The abrupt cessation of successful pregnancies at or just before 40 years of age is very much in keeping with the observed norms in developing societies and may be ascribable to the rapid decline in fecundability in the years immediately preceding menopause (Campbell & Wood, 1988).

Maternal age profiles at first liveborn pregnancy, subdivided by religion,

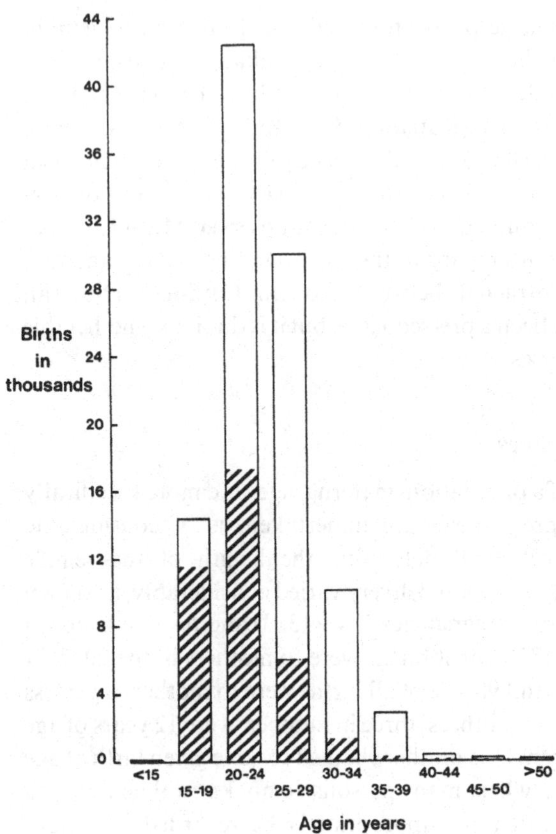

Figure 6.1. Maternal age profiles for all liveborn pregnancies (total columns, $n = 99\,323$) and for first liveborn pregnancies only (hatched areas, $n = 35\,173$).

are presented in Figure 6.2. Muslims tended to initiate fertility earliest: by 20 years of age 38.8% of Muslim women had given birth to their first child compared to 31.7% of Hindus and 21.0% of Christians. The age profile of Christian females at first birth was the most symmetrical, the oldest primiparous Christian female being 37 years of age.

As seen in Figure 6.3, maternal age at first liveborn pregnancy also was strongly influenced by consanguinity, with an inverse relationship between maternal age and level of inbreeding. For example, of the first liveborn pregnancies of women under 20 years old the percentages observed in each consanguinity class were nonconsanguineous (28.7%), second cousin (37.2%), first cousin (39.4%) and uncle–niece (42.0%). Thus both religion and consanguinity might be expected to influence fertility *via* differentials in maternal reproductive span.

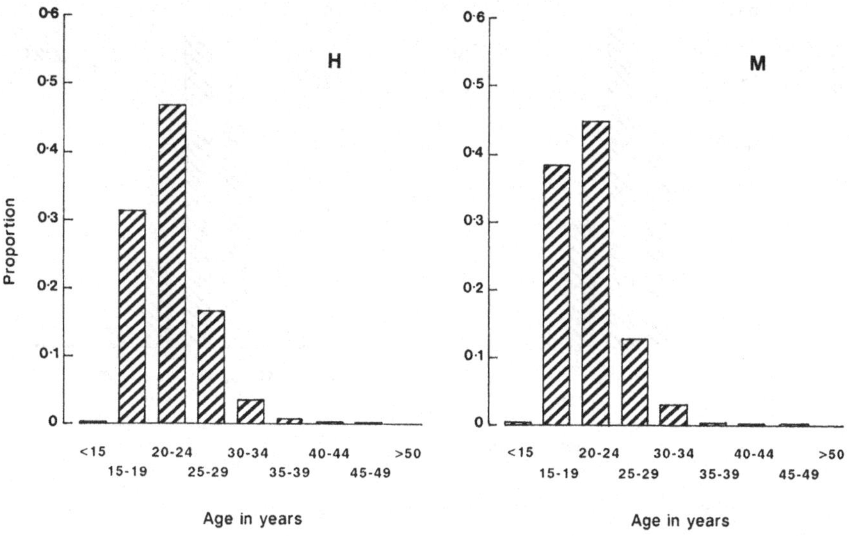

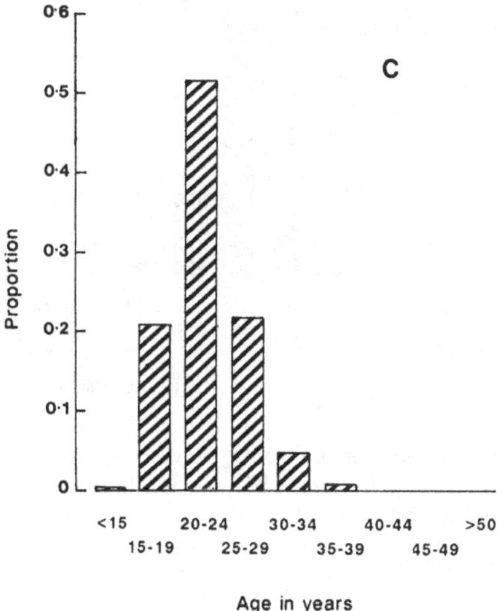

Figure 6.2. Maternal ages at first liveborn pregnancy by religion: Hindu (H, $n = 29\,823$), Muslim (M, $n = 5134$) and Christian (C, $n = 1427$).

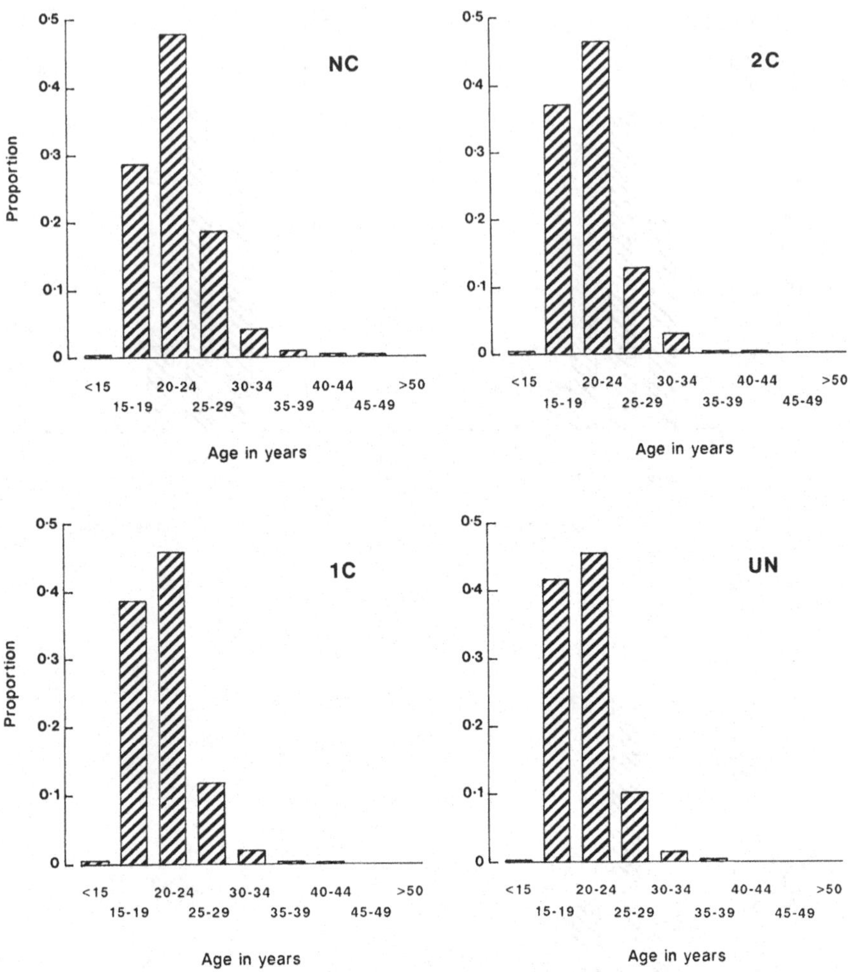

Figure 6.3. Maternal ages at first liveborn pregnancy by consanguinity: Non-consanguineous (NC, $n = 23\,368$), second cousin (2C, $n = 586$), first cousin (1C, $n = 3902$) and uncle–niece (UN, $n = 5809$).

Specific aspects of fertility

(1) Multiple birth rates

Of the 99 323 pregnancies delivered, there were 695 pairs of twins, 9 sets of triplets and 1 set of quadruplets. As previously noted (Bittles *et al.*, 1988), neither religion nor consanguinity has a significant effect on the twinning rate in the population which, at 7.0 per thousand maternities, was in the lower range reported for human populations. In agreement with the findings of Bulmer (1970), the mean maternal age of twin pregnancies was older, 25.2 years SD 4.70, than in singleton pregnancies, 23.9 years SD 4.46.

(2) *Secondary sex ratio*

The sex ratio at birth, 0.5217, was higher than in comparable major human populations but again it was unaffected by religion or consanguinity. The apparent elevation in the proportion of male births in Karnataka is somewhat puzzling as it could not be ascribed to any of the mechanisms commonly proposed to influence the secondary sex ratio (James, 1987). Equally, there is no evidence of sex-selective foeticide which has been reported in some North Indian states (Jeffrey *et al.*, 1984). The simplest and probably most convincing explanation for the observation of greater than expected numbers of male births is that parents limit their families once the desired number of male children has been achieved, thus artificially biasing the secondary sex ratio and augmenting the pre-existing excess of males at birth.

(3) *Religion and fertility*

The numbers of liveborn and living children in each of the three main religious groups are shown for all pregnancies in Table 6.3. As 36.6% of the pregnancies were firstborn deliveries i.e. the category one liveborn, one living child, the numbers observed are biased towards a smaller mean family size but this caveat applies to all religions and consanguinity classes. Both in terms of mean numbers of liveborn and living children Muslims had larger family sizes than Hindus or Christians and the variances associated with each parameter also were largest for Muslims. Tested by one-way ANOVA, the difference in mean values between the three religious groups just failed to attain statistical significance at the 0.05 level for liveborns but was significant ($P<0.05$) for living children. No significant difference was seen between religions in terms of survivorship, with relative proportions calculated for Hindus, Muslims and Christians of 0.9631, 0.9688 and 0.9676. Therefore the greater net Muslim fertility appears to be independent of any childhood mortality effect. Again it should be noted that these values are influenced by the high percentage of first liveborn pregnancies in the study population.

Comparable figures for mean numbers of liveborn and living children born to all ever-married women in Karnataka are shown in Table 6.4, based on the 1981 Census of India (Government of India, 1986). In these

Table 6.3. *Numbers of liveborn and living children by religion*

	Number of liveborns		*Number of living children*	
Religion	*Mean*	*Variance*	*Mean*	*Variance*
Hindu	2.17	0.38	2.09	0.35
Muslim	2.56	0.88	2.48	0.83
Christian	2.16	0.48	2.09	0.41

Table 6.4. *Mean numbers of liveborn and living children born to all ever-married women in Karnataka, by religion*

	Urban		*Rural*	
	Liveborn	*Living*	*Liveborn*	*Living*
Hindu	3.25	2.77	3.45	2.80
Muslim	3.86	3.31	3.91	3.20
Christian	3.44	3.03	4.15	3.53

data the greater overall fertility of Muslim women is confirmed for urban families, but in rural communities the small Christian group has the largest mean family sizes.

(4) *Consanguinity and fertility*

The relationships between consanguinity and mean numbers of liveborn and living children are shown in Table 6.5. No statistically significant trend with increasing consanguinity was apparent in either parameter but the numbers of both liveborn and living children were smallest in nonconsanguineous marriages. The proportion of survivors was comparable across all five consanguinity classes: nonconsanguineous (0.9640), beyond second cousin (0.9734), second cousin (0.9668), first cousin (0.9595) and uncle–niece (0.9571). However, there was a correlation between consanguinity and multiple deaths in families (Bittles *et al.*, in preparation). For example, in families with no deaths the offspring of uncle–niece and first cousin marriages accounted for 29.3% of the total, with two deaths in the family they totalled 31.5% and with four and six deaths the levels increased to 43.3% and 41.4% respectively. To place these figures in perspective; families with one or more deaths accounted for only 5.7% of the 99 323

Table 6.5. *Numbers of liveborn and living children by consanguinity*

	Number of liveborns				*Number of living children*			
Marital type	*Max.*	*Min.*	*Mean*	*Variance*	*Max.*	*Min.*	*Mean*	*Variance*
Nonconsanguineous	15	1	2.20	0.41	15	1	2.12	0.38
Beyond second cousin	17	1	2.27	0.44	10	1	2.21	0.41
Second cousin	10	1	2.32	0.59	10	1	2.24	0.53
First cousin	14	1	2.35	0.66	14	1	2.25	0.56
Uncle–niece	11	1	2.28	0.61	10	1	2.18	0.53

pregnancies studied and, even if firstborn deliveries are excluded, this figure remains at less than 10% of the total.

Discussion

The present case study indicates that religion and consanguinity play a significant and inter-related role in determining the overall level of fertility exhibited by this South Indian population. High levels of fertility are characteristic of Muslim populations world-wide (Nagi, 1983; Ahmad, 1985; Modell, 1988). In Karnataka, Muslim marriages are seen to have early initiation of pregnancies (Figure 6.2) and successful pregnancies continuing into the 40+ years age group. Thus, there is evidence that their reproductive span is greater than that of Hindus and, more especially, of Christians. Additional factors which require consideration in assessing the fertility of Karnatakan Muslims include their poor mean socio-economic and educational status. In combination with the ambiguity which surrounds Islamic doctrinal attitudes to contraception, it seems less likely that they practise birth control and their religious beliefs also generally would preclude acceptance of therapeutic abortion (Nagi, 1983). Finally, there is the questionable effect of polygyny on the fertility of the Muslim community. Data on this subject frequently are conflicting (Sembajwe, 1979) and currently it is difficult either to quantify the numbers of such marriages among Karnatakan Muslims or to estimate their overall effect on family size.

In view of the frequency of consanguineous marriages world-wide, it is somewhat puzzling that greater attention has not been paid to the role of inbreeding in determining fertility. Studies in Japan have shown that consanguineous marriages generally have lower rates of sterility and greater numbers of liveborn pregnancies (Schull *et al.*, 1962; 1970). Yet surveys on fertility conducted in countries such as Pakistan (Ahmad, 1985), where between 38 and 39% of marriages currently are consanguineous (Shami *et al.*, 1989), have failed even to acknowledge the existence of inbreeding in the community. In Karnataka it is clear that consanguinity is strongly related to age at first liveborn pregnancy and hence to reproductive span (Figure 6.3). The relationship between the two parameters is scarcely surprising. Both Hindus and Muslims preferentially advocate marriage between close relatives and the requisite marriage arrangements can readily be made within the family circle. Among South Indian Hindus various reasons are advanced for the popularity of consanguineous marriages, but mainly they centre on benefit perceived in terms of the enhanced social stability which is believed to result from marriage to a relative, and the maintenance of family property (Dronamraju & Meera Khan, 1963; Reid, 1973). Certainly, the fact that marriage to a close relative results in minimal dowry or bride wealth payments (Hann, 1985b; Govinda Reddy, 1988) must be a strong economic incentive in most communities.

The future incidence of consanguinity in South India and its influence on family size is difficult to assess. A decrease in consanguineous unions has been noted in Andhra Pradesh (Dronamraju, 1964) and southern Karnataka (Caldwell *et al.*, 1983) but there is no evidence of a decline in their popularity during the last decade in Bangalore and Mysore (Bittles *et al.*, 1987). However, given the move to smaller family sizes, the later age of women at marriage and greater population mobility, a future decline at least in uncle–niece marriages would appear to be inevitable (Radha Rama Devi *et al.*, 1982).

The degree to which parental fertility is influenced by child mortality and vice versa is problematical. This question becomes especially important in consanguineous marriages, since offspring may exhibit high rates of disorders caused by the expression of rare recessive genes inherited from a common ancestor(s). In Japan it was concluded that the greater fertility of consanguineous marriages probably resulted from reproductive compensation, with the replacement of children dying at a young age (Schull *et al.*, 1970; Schull & Neel, 1972). As shown in Table 6.5, in Karnataka there is no evidence of a significant effect of consanguinity on overall survivorship and so the slightly larger numbers of children born to consanguineous families can be most readily be explained in terms of younger initiation of pregnancies. Nevertheless, the over-representation of uncle–niece and first cousin offspring in families with multiple deaths suggests that, in certain cases, consanguineous families or groups may be at high risk from genetic disorders. Support for this assumption is provided by the findings of an earlier clinical and laboratory-based investigation of children referred with essentially non-specific symptoms (Radha Rama Devi *et al.*, 1987). With the continuing decline of nutritional and infectious diseases as causes of childhood morbidity and mortality in the region, it can be predicted that inherited disorders will increase proportionately in significance in future generations.

A final factor to be considered in discussing preferred marital patterns and fertility in Karnataka is the sex ratio of the general population. Increases in the tertiary sex ratio were reported in decennial censuses conducted from 1901 and by 1981 the values for urban Bangalore and Mysore were 0.5615 and 0.5335 with a state-wide mean value of 0.5190 (Bittles *et al.*, 1988). Somewhat surprisingly the growing disequilibrium in sex ratio has not adversely affected prospects of male marriage, because the overall surplus of males can be accommodated within accepted societal norms by marriage to younger females. For example, in Bangalore the mean spousal age difference currently is 7.9 years. A strategy of this nature is made possible by rapid population increase, which ensures ever-larger cohorts of marriageable females (Krishnamoorthy, 1977). Hence it has been postulated that, at least in the short term, a surplus of marriageable females has effectively been created, which in turn may explain the recent transition from bride wealth to dowry. In the current generation the situation has been exacerbated by a narrowing of spousal age differences due to an increase in female age at marriage with no equivalent adjustment among males (Krishnamoorthy, 1977; Kulkarni *et al.*, 1986), to the extent that it has been suggested some women may never marry (Caldwell *et al.*, 1983). However, if the trend towards an increasing preponderance of males continues but at the same time birth rates decline, then in future generations the position may reverse and a proportion of males be destined for obligatory bachelorhood. Future studies in all of these areas should prove to be of considerable interest, not least in their implications with respect to fertility of the population.

References

Ahmad, S. (1985). Factors affecting fertility in four Muslim populations: a multivariate analysis. *Journal of Biosocial Sciences*, **17**, 305–16.

Ali, S. G. M. (1968). Inbreeding and endogamy in Kerala (India). *Acta Genetica et Statistica Medica*, **18**, 369–79.

Appaji Rao, N., Radha Rama Devi, A., Savithri, H. S., Venkat Rao, S. & Bittles, A. H. (1988). Neonatal screening for amino acidaemias in Karnataka, South India. *Clinical Genetics*, **34**, 60–3.

Basu, A. M. (1987). Household influences on childhood mortality: evidence from historical and recent mortality trends. *Social Biology*, **34**, 187–205.

Basu, S. K. (1975). Effect of consanguinity among North Indian Muslims. *Journal of Population Research*, **2**, 57–68.

Bittles, A. H., Radha Rama Devi, A. & Appaji Rao, N. (1988). Consanguinity, twinning and secondary sex ratio in the population of Karnataka, South India. *Annals of Human Biology*. **15**, 455–60.

Bittles, A. H., Radha Rama Devi, A., Savithri, H. S. & Appaji Rao, N. (1987). Consanguineous marriage and postnatal mortality in Karnataka, South India. *Man (N.S.)*, **22**, 736–45.

Bulmer. M. B. (1970). *The Biology of Twinning in Man*, p. 74. Oxford: Clarendon Press.

Caldwell, J. C., Reddy, P. H. & Caldwell, P. (1983). The causes of marriage change in South India. *Population Studies*. **37**, 343–61.

Campbell, K. L. and Wood, J. W. (1988). Fertility in traditional societies. In: *Natural Human Fertility: Social and Biological Determinants*, ed. P. Diggory, M. Potts and S. Teper, p. 39. London: Macmillan Press.

Dronamraju, K. R. (1964). Mating systems of the Andhra Pradesh people. *Cold Spring Harbor Symposia on Quantitative Biology*, **29**, 81–4.

Dronamraju, K. R. & Meera Khan, P. (1963). The frequency and effects of consanguineous marriages in Andhra Pradesh. *Journal of Genetics*, **58**, 387–401.

Government of India (1986). *Census of India*, 1981. Series-9 Karnataka, Part VI, Fertility, B. K. Das (ed.). Mysore: Government of India Text Book Press.

Govinda Reddy, P. (1988). Consanguineous marriages and marriage payment: a study among three South Indian caste groups. *Annals of Human Biology*, **15**, 263–68.

Hann, K. L. (1985a). Inbreeding and fertility in a South Indian population. *Annals of Human Biology*, **12**, 267–74.

Hann, K. L. (1985b). The incidence of relation marriage in Karnataka, South India. *South Asia Research*, **5**, 59–72.

James, W. H. (1987). The human sex ratio. Part 1: a review of the literature. *Human Biology*, **59**, 721–52.

Jeffery, R., Jeffery, P. & Lyon, A. (1984). Female infanticide and amniocentesis. *Social Science and Medicine*, **19**, 1207–12.

Kapadia, K. M. (1958). *Marriage and Family in India*, 2nd edn., p. 117. Calcutta: Oxford University Press.

Krishnamoorthy, S. (1977). An enquiry into the effect of the disequilibrium in sex ratio on marriage in India. *Demography India*, **6**, 182–92.

Kulkarni, P. M., Savanur, L. R. & Gokhale, C. V. (1986). Increase in age at marriage in rural Karnataka: evidence from a repeat survey. *Demography India*, **15**, 149–63.

Kumar, S., Pai, R. A. & Swaminathan, M. S. (1967). Consanguineous marriages and the genetic load due to lethal genes in Kerala. *Annals of Human Genetics*, **31**, 141–45.

Modell, B. (1988). Interpretation and application of demographic data. *Biology and Society*, **5**, 27–32.

Nagi, M. H. (1983). Trends in Moslem fertility and the application of the demographic transition model. *Social Biology*, **30**, 245–62.

Radha Rama Devi, A., Appaji Rao, N. & Bittles, A. H. (1981). Consanguinity, fecundity and post-natal mortality in Karnataka. *Annals of Human Biology*, **8**, 469–72.

Radha Rama Devi, A., Appaji Rao, N. & Bittles, A. H. (1982). Inbreeding in the state of Karnataka, south India. *Human Heredity*, **32**, 8–10.

Radha Rama Devi, A., Appaji Rao, N. & Bittles, A. H. (1987). Inbreeding and the incidence of childhood genetic disorders in Karnataka, South India. *Journal of Medical Genetics*, **24**, 362–5.

Rao, P. S. S. (1983). Religion and the intensity of inbreeding in Tamil Nadu, south India. *Social Biology*, **30**, 413–22.

Rao, P. S. S. & Inbaraj, S. G. (1977). Inbreeding in Tamil Nadu, south India. *Social Biology*, **24**, 281–8.

Reid, R. M. (1973). Social structure and inbreeding in a south Indian caste. In: *Genetic Structure of Populations*, (ed.) N. E. Morton, p. 92. Honolulu: University of Hawaii Press.

Sanghvi, L. D. (1966). Inbreeding in India. *Eugenics Quarterly*, **13**, 291–301.

Schull, W. J., Furusho, T., Yamamoto, M., Nagano, H. & Komatsu, I. (1970). The effect of parental consanguinity and inbreeding in Hirado, Japan. IV. Fertility and reproductive compensation. *Humangenetik*, **9**, 294–315.

Schull, W. J. & Neel, J. V. (1972). The effects of parental consanguinity and inbreeding in Hirado, Japan. V. Summary and interpretation. *American Journal of Human Genetics*, **24**, 425–53.

Schull, W. J., Yanase, T. & Nemoto, H. (1962). Kuroshima: the impact of religion on an island's genetic heritage. *Human Biology*, **34**, 271–98.

Sembajwe, I. (1979). Effect of age at first marriage, number of wives, and type of marital union on fertility. *Journal of Biosocial Science*, **11**, 341–51.

Shami, S. A., Schmitt, L. H. & Bittles, A. H. (1989). Consanguinity-related prenatal and postnatal mortality of the populations of seven Pakistani Punjab cities. *Journal of Medical Genetics*, **26**, 267–71.

7 *Resources and the fertility transition in the countryside of England and Wales*

P. R. ANDREW HINDE

Introduction

Between 1871 and 1931, the average completed family size in England and Wales declined from around six or seven children to around two or three children. This decline in fertility was unprecedented; it was universal in that it ultimately affected all sections of society; and, so far as we can tell at present, it was irreversible. It was a social change of monumental importance and has been aptly labelled the *fertility transition*. Together with the decline in mortality which took place after 1750, it forms the well-known phenomenon of the *demographic transition*.

The fertility transition in England and Wales had two important characteristics. First, it was due almost entirely to a decline in fertility within marriage: the propensity of the population to marry hardly changed (Figure 7.1). Second, it involved a shift from uncontrolled to controlled fertility as married couples began to take conscious decisions to limit the number of children they had. It is argued (for example by Knodel, 1988: pp. 247–349) that this shift involved the gradual adoption by the population of '*stopping*' *behaviour*, whereby couples began to use fertility control methods once they had produced their desired number of children.

Figure 7.1. Overall fertility, marital fertility and proportions married: England and Wales, 1851–1931.

Notes. The graph plots the values of the fertility indices devised by Coale (1967). These indices are useful in the analysis of fertility in nineteenth century England and Wales because age-specific fertility rates are not available. Two of the indices, I_f and I_g, compare the observed number of births in a given year in the population under analysis (which we will call population A) with the number that would have been produced by a population having that age structure if the age-specific fertility rates of the women in population A had been equal to the highest on record, namely those of married Hutterite women in 1921–30. In other words, they are a method of comparing the fertility of two populations adjusting for differences in their

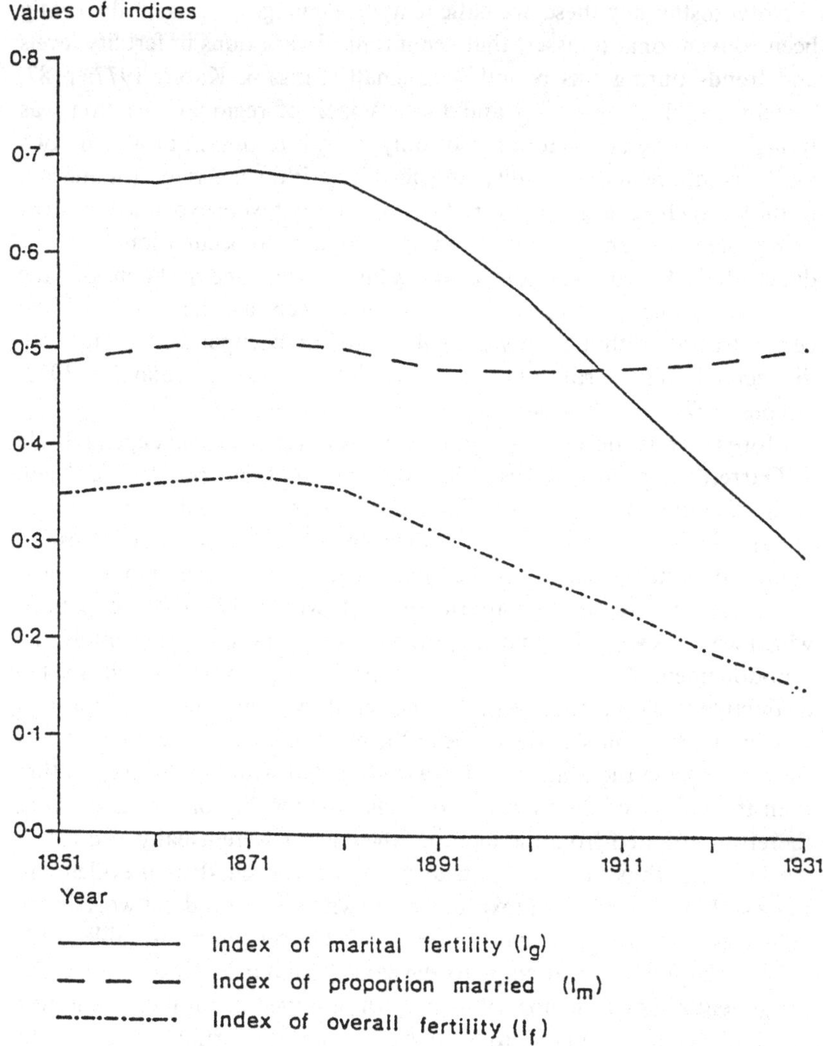

respective age structures by *indirect standardization*. The *index of overall fertility*, I_f, therefore, is the ratio of the total number of births observed in population A (in the given year) to the number that the women in population A would have produced had their age-specific fertility rates been the same as the Hutterites'. The *index of marital fertility*, I_g, is defined in a similar way, but uses only the legitimate births and the married women in population A. The *index of nuptiality*, I_m, for population A is a weighted average of the extent to which the proportion of women in each childbearing age-group in that population who are married falls below 100%, where the weights are the age-specific fertility rates of married Hutterite women in 1921–30. *Source*: Teitelbaum (1984: Tables 4.2, 5.3 and 6.2, pp. 79, 101 and 117).

Notwithstanding these dramatic temporal changes, it has until recently been conventional to assert that geographical variations in fertility levels and trends during this period were small (Brass & Kabir, 1977: p. 87; Teitelbaum, 1984: pp. 88–9 and 149). A lack of regional variation was thought to have characterized not only overall fertility, but also its two major components: nuptiality and fertility within marriage (or marital fertility). Such geographical variations as did exist were explained away as being regional manifestations of more fundamental occupational fertility differentials. For example, the relatively high fertility and the lateness of the fertility decline in county Durham were seen as the result of the concentration in that county of coal miners, an occupational group with distinctively high fertility and a late fertility decline (Friedlander, 1973; Haines, 1979: pp. 155–204).

More recently, however, this interpretation has been challenged (Hinde & Garrett, in press; Woods, 1982: pp. 112–24; 1986: pp. 29–32). These authors maintain, first, that regional differences in nuptiality and fertility patterns in nineteenth century England were more pervasive and persistent than had hitherto been thought and, second, that is was not so much occupation *per se*, as the patterns of work within different occupations which were associated with demographic variation. Thus, for example, the well-documented low fertility of textile workers was related to the distinctive work pattern in the textile industry, which involved married women working outside the home in the mills. It was the fact that women worked outside the home which was associated with low fertility, rather than the nature of their occupation. The low fertility of textile workers, therefore, resulted from the fact that their wives were usually also textile workers, and thus employed outside the home. Indeed, there is evidence to suggest that the fertility of textile workers whose wives did not work in the mills was no lower than average (Hinde & Garrett, in press, table 11).

This chapter is an attempt to develop a similar explanation for one prominent regional demographic pattern observed in rural communities, the tendency for nuptiality to be lower, and marital fertility higher, in the north and west of England than in the south-east. The account essayed here relates this pattern to differences in the structure of the agricultural work force in the two regions. It will be suggested that contrasting work patterns caused the economic dynamics of households to operate differently in the two areas, and that this resulted in differences in fertility levels and perhaps in the timing of the onset of voluntary fertility control. Thus the interpretation offered explicitly relates household resources to fertility trends.

The next two sections describe, respectively, the geography of nuptiality, marital fertility and the structure of the agricultural labour force in 1861;

and changes in these factors over the subsequent 50 years, using empirical evidence based on census and civil registration data. I then discuss theoretically why agricultural work patterns should be associated with those in marriage and fertility. Finally, we link these theoretical arguments to the empirical evidence. In order to keep the chapter to a reasonable length, some of the arguments have been reduced to a simple (and possibly a simplistic) form; and many qualifications and caveats have been omitted. What follows, therefore, resembles a 'potted version' of what is really a much more complex story. Parts of this chapter are also undoubtedly speculative. Of course speculation is necessary in historical demography, since empirical evidence is frequently lacking. I have, though, presented some of the arguments in a deliberately speculative form, hoping that those who disagree with my interpretation will be encouraged to offer their own alternatives, and so advance our knowledge of the fertility transition and its causes.

The geography of nuptiality, marital fertility and agricultural employment in 1861

Nuptiality and marital fertility

Measures of the level of nuptiality and marital fertility in 1861 in each of the 600 or so registration districts of England and Wales have been calculated. The resulting figures can be used to draw maps of the geography of marriage and marital fertility at that date, omitting those registration districts with a predominantly non-agricultural population (Figures 7.2 and 7.3).

The map of nuptiality levels (Figure 7.2) shows clearly the contrast between an area to the north and west of a line running roughly from Exeter in Devon to Scarborough in Yorkshire, and the rest of England and Wales. Because we have had to group the values of our measure of nuptiality into a small number of categories in order to draw the map, the latter gives a rather misleading impression of the abruptness of the division between the two zones. In fact nuptiality seems to have decreased rather continuously from the south-east to the north-west, with the highest values being found around the Thames estuary and the lowest values in west Wales and along the Scottish border.

The geography of marital fertility shows a similar pattern (Figure 7.3), with higher values being found in those northern and western areas where nuptiality was relatively low, and lower values in the south-east. The contrast does not seem from the map to be as acute as that for nuptiality, but if we again use the Exeter–Scarborough line to divide England and Wales into two zones, it is quite impressive in numerical terms (Table 7.1).

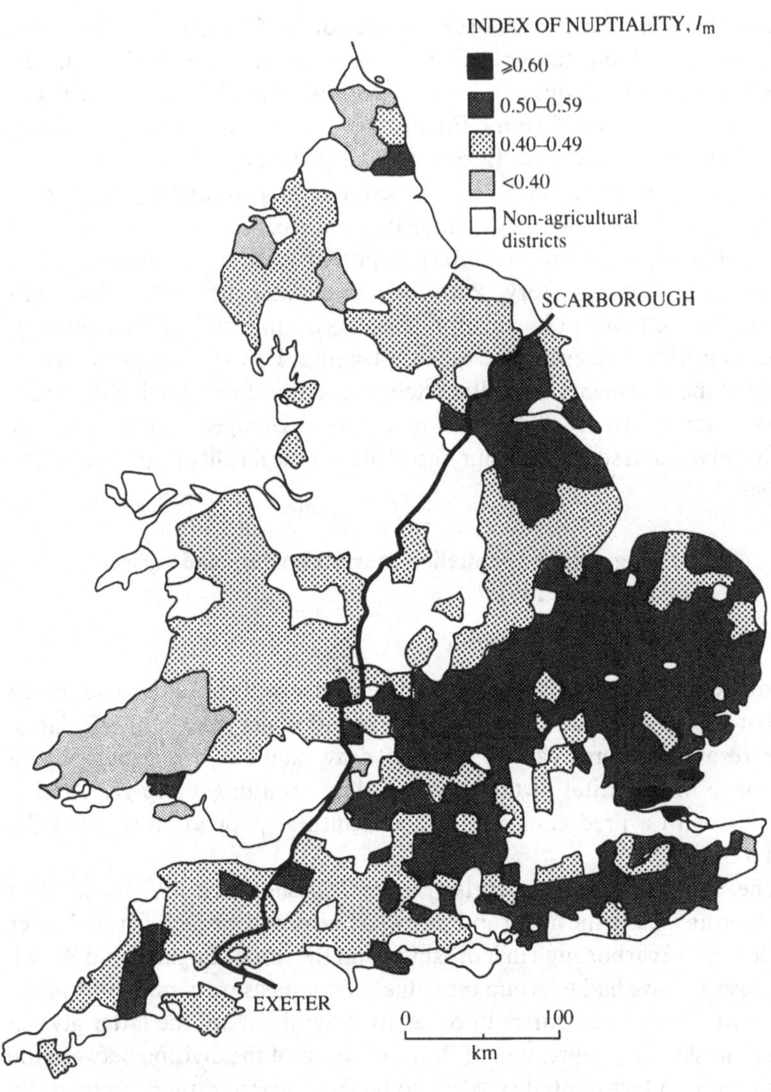

Figure 7.2. Nuptiality in the agricultural registration districts of England and Wales, 1861.

Notes. The index I_m is defined in the notes to Figure 7.1. 'Agricultural' registration districts are defined to be those in which at least 35% of the adult male population in 1871 was employed in the agricultural sector. The year 1871 is used in this definition (rather than 1861) because the Census tabulations in 1861 were arranged in such a way that the calculation of such a percentage for that year is very onerous.

Source: Data supplied by Robert Woods; Census of England and Wales (1871, various pp.).

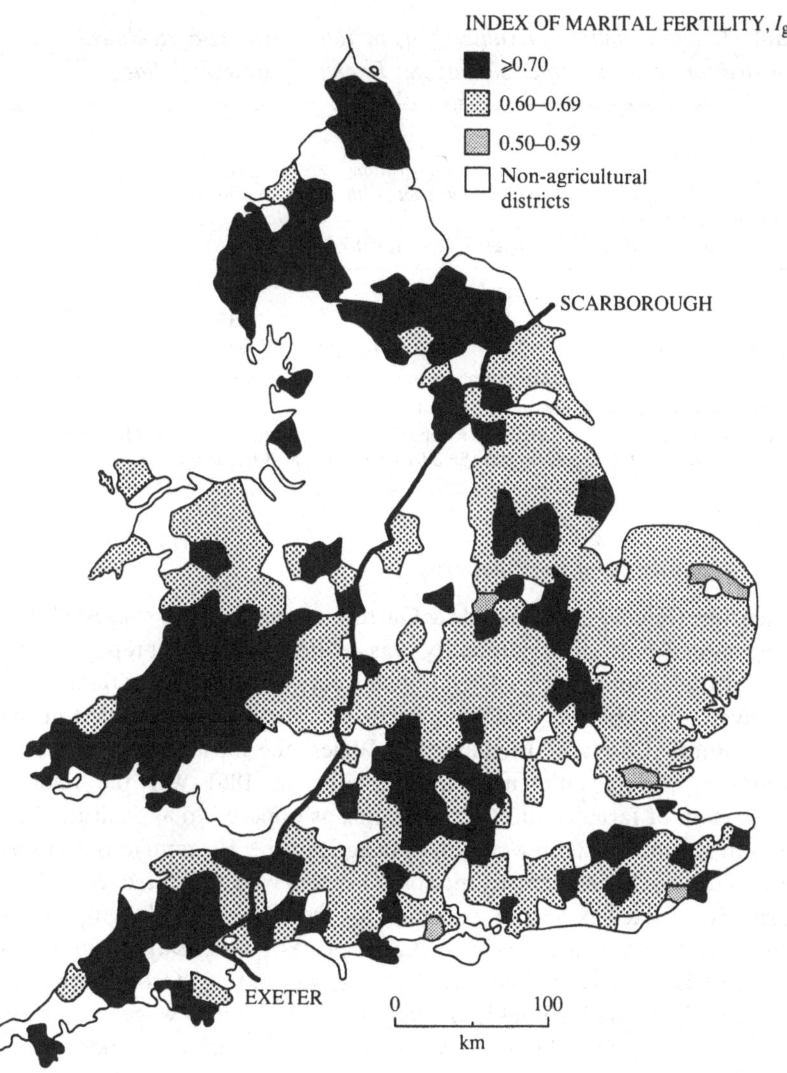

Figure 7.3. Marital fertility in the agricultural registration districts of England and Wales, 1861.
Notes. The index I_g is defined in the notes to Figure 7.1. 'Agricultural' registration districts are defined in the notes to Figure 7.2. *Source*: Data supplied by Robert Woods; Census of England and Wales (1871, various pp.).

Table 7.1. *Distribution of values of I_g in 1861 amongst agricultural registration districts either side of the Exeter–Scarborough line*

Position in relation to Exeter–Scarborough line	*Number of registration districts with* $I_g \geq 0.7$	$I_g < 0.7$	*Total number of registration districts*
North-west	72	32	104
South-east	48	189	237
Totals	120	221	341

Notes: The index I_g is defined in the notes to Figure 7.1.
Agricultural registration districts are defined in the notes to Figure 7.2. The association between zone and I_g is significant at the 0.01 level when a χ^2 test is applied.

Agricultural work patterns

Elsewhere (Hinde, 1985b; Hinde & Garrett, in press), I have suggested that the regional pattern of nuptiality was related to the geography of the agricultural labour supply. A relationship between these two variables was discovered by Michael Anderson (1976, pp. 74–6) who found that in the agricultural regions of England and Wales, the most important single factor explaining differences in nuptiality in 1861 was the relative proportion of farmers and farm servants, as opposed to agricultural day labourers, among all those employed in agriculture. The term *farm servants* is used to refer to young men (and some women) who were hired to work on farms for a fixed period in exchange for free board and lodging and a stipend. Farm servants thus lived at their place of work. *Day labourers*, on the other hand, lived in their own cottages, walked to and from work each day, and were paid a weekly or daily (or even hourly) wage. Anderson found that low levels of nuptiality were associated with high proportions of the agricultural work force being provided by farm servants, and vice versa.

The geography of farm service in mid-nineteenth century England has been examined by Ann Kussmaul (1981: pp. 19–22). She drew a map showing farm servants as a percentage of the agricultural work force in each county in 1851. Here, we present a similar map for 1871 (Figure 7.4). The contrast between the area to the south-east of a line joining the Wash to the Exe, where farm service was almost extinct, and the rest of the country, where it was still important, is striking. There is a close (although not exact) correspondence between the geography of agricultural work patterns and that of nuptiality (Figure 7.2).

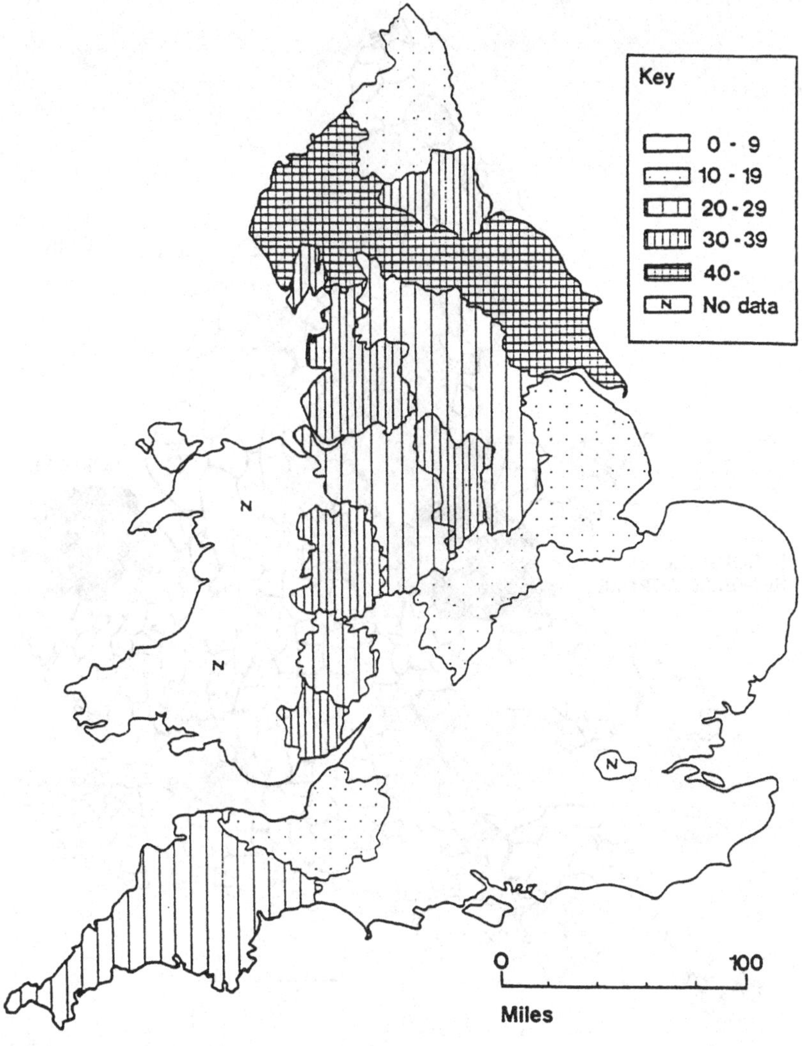

Figure 7.4. Percentage of all farm workers described as farm servants: England and Wales, 1871. *Source*: Census of England and Wales (1871).

The evolution of nuptiality, marital fertility and work patterns between 1861 and 1911

We can examine how nuptiality and marital fertility patterns in the two zones of England and Wales changed over the 50 years after 1861 by examining their evolution in two groups of registration districts. One of these groups is in the county of Norfolk, in eastern England; the

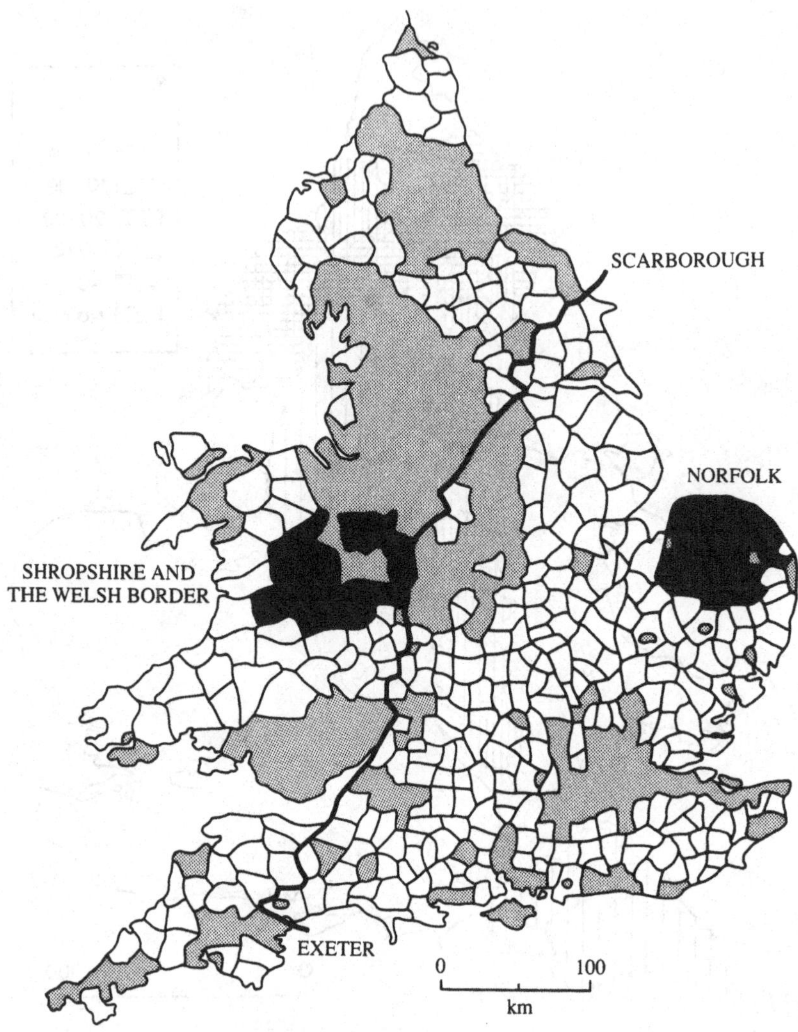

Figure 7.5. Location of registration districts for which nuptiality and marital fertility are plotted in Figure 7.6.

Figure 7.6. Nuptiality and marital fertility in selected registration districts in Norfolk, and in Shropshire and the Welsh Border, 1861, 1891 and 1911.
Notes. The indices I_m and I_g are defined in the notes to Figure 7.1. The registration districts for which they have been plotted in this figure are; in Norfolk – Aylsham, Blofield, Docking, Depwade, Downham, Erpingham, Forehoe, Freebridge Lynn, Guiltcross, Henstead, Mitford, St Faith's, Smallburgh (1861 and 1891 only), Swaffham, Thetford, Walsingham and Wayland; and in Shropshire and the Welsh Border – Bridgnorth, Church Stretton, Clun, Ellesmere, Forden, Llanfyllin, Ludlow, Market Drayton, Knighton, Newport, Newtown, Oswestry, Shiffnal, Wem, and Whitchurch. The locations of these registration districts are shown in Figure 7.5.
Source: Data supplied by Robert Woods.

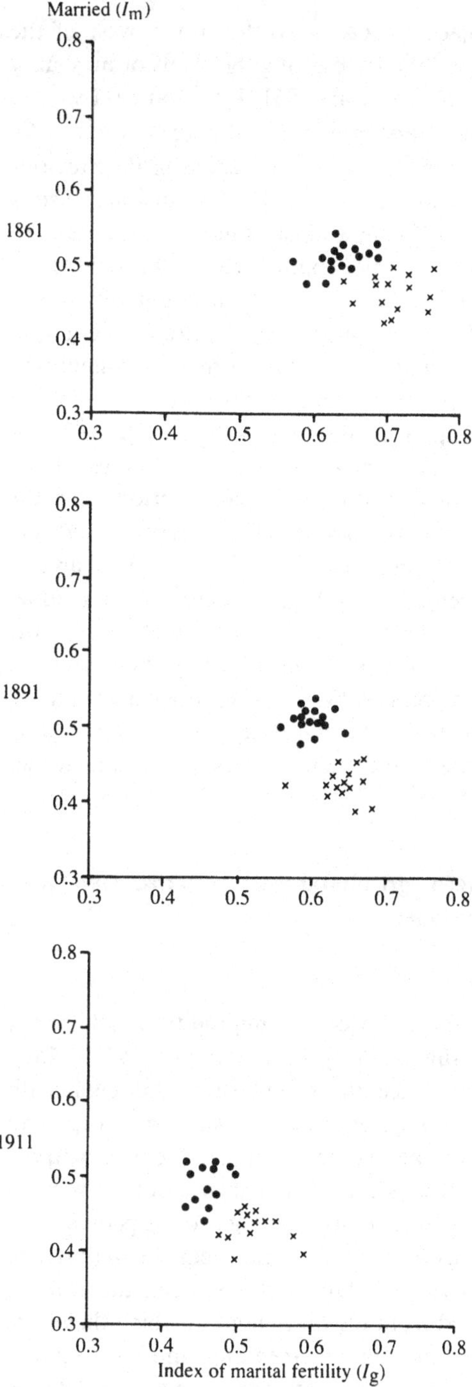
Married (I_m)
0.8
0.7
0.6
1861
0.5
0.4
0.3
0.3 0.4 0.5 0.6 0.7 0.8
1891
1911
Index of marital fertility (I_g)
• Norfolk
× Shropshire and the Welsh border

other is in Shropshire and adjacent areas of Wales, north-west of the Exeter–Scarborough line (Figure 7.5). By plotting the levels of nuptiality against those of marital fertility for the years 1861, 1891 and 1911 we can examine the course of the fertility transition in the two areas (Figure 7.6).

In 1861, as we might expect from the results presented in the previous section, the two groups of registration districts fall into two distinct clusters on the graph. Between 1861 and 1911 these clusters move horizontally to the left, but their vertical position does not change much. This reflects the fact that during this period marital fertility declined whilst nuptiality stayed much the same (Figure 7.1). The most interesting feature of the graph, however, is that the *relative* position of the two clusters remains roughly the same in 1911 as it had been in 1861. There seems to have been no reduction in geographical differences in nuptiality and marital fertility by 1911.

It is, unfortunately, much more difficult to obtain estimates of the evolution of agricultural work patterns during the same period, since the censuses after 1881 do not distinguish between farm servants and day labourers. Data from the original census returns books for a small area in Shropshire suggest that between 1861 and 1881 there was a gradual attenuation of the prevalence of farm service. During that period, the percentage of all male agricultural workers 'living in' fell by one-third from 34.7 to 23.3 (Hinde & Garrett, in press, Table 4). Such evidence as there is, therefore, suggests that the extinction of farm service as a way of providing agricultural labour in these northern and western areas was gradual rather than abrupt, at least until 1881.

The relationships between agricultural work patterns, nuptiality and fertility within marriage.

Work patterns and nuptiality

Why should the prevalence of farm service and nuptiality be related? A plausible mechanism has been suggested by John Hajnal (1982: p. 452). Hajnal states that during the sixteenth, seventeenth and eighteenth centuries the household formation system in north-western Europe had three characteristics. First, marriage occurred, on the average, at quite advanced ages for both sexes. Second, after marriage a couple were in charge of their own household. Third, before marriage young people often circulated between households as servants (in rural areas the majority of these would have been farm servants). Under this system, therefore, a couple moved when they married from the household in which they were previously living into their own household. Spending time as a servant provided young men and women with the opportunity to accumulate the resources necessary to form the new household. Because saving up in this way took time, marriage was delayed for both sexes. As a result north-western Europe was characterized by a high prevalence of farm service and low nuptiality.

During the first half of the nineteenth century, however, farm service declined almost to the point of extinction in south-east England (Snell, 1985: pp. 67–103). This meant that the possibility of working as a farm servant in order to accumulate the wherewithal to marry disappeared in that area. The need to form one's own household upon marriage, however, remained as strong as ever. For example, in central Norfolk during the second half of the nineteenth century (where there were almost no farm servants) only about one per cent of households contained more than one married couple (Hinde, 1985a: Table 9.3, p. 395).

Work patterns and marital fertility

If farm service was not an option, how were couples marrying during the second half of the nineteenth century in south-east England to gain the resources necessary to form their own household? For young women, domestic service was still a possibility; but young men could only save by remaining in their parents' households whilst working as day labourers on a nearby farm. The fact that by 1861 nuptiality in the south and east was higher than in the north and west suggests either that this was a more effective method than working as a servant (in that it permitted faster saving) or that it was less desirable from the point of view of the young men, and encouraged them to marry earlier with a smaller resource base as a means of escape. The first of these seems implausible, since, if it were true it is difficult to explain the great prevalence of the institution of service throughout north-western Europe during the sixteenth, seventeenth and eighteenth centuries. To summarize, therefore, we suggest that young people in the south-east of England during the second half of the nineteenth century married with fewer resources than their peers in the north and west.

The argument can be taken further by suggesting that the households of those who married with fewer resources would subsequently have been more vulnerable to economic strain than would those of their better-endowed peers. It has long been maintained that economic strain may influence demographic behaviour (Davis, 1963: pp. 349–50; Friedlander, 1983: pp. 249–51). The resource base of working-class rural households in Victorian England was always fairly precarious, and economic 'crises' (occurring, for example, when the breadwinner was ill or when there were large numbers of dependent children in the household) must have been a recurrent feature of the lives of many families. One possible response to economic strain is a reduction of fertility through voluntary fertility control. The link between work patterns and fertility is completed, therefore, by suggesting that the greater economic vulnerability of households in the south and east of England (relative to those in the north and west) meant they were more likely to encounter economic 'crises' during their marriages and to reduce their fertility in response.

Discussion

Let us return to the geographical pattern of marital fertility in 1861 (Figure 7.3). If we are to account for the marital fertility differentials in 1861 by this argument, it seems that we are constrained to believe that couples in mid-nineteenth century rural England had access to methods of deliberate fertility control within marriage. For only if this is so can we explain the lower marital fertility of south-eastern couples in terms of their greater readiness to reduce their fertility in the face of correspondingly greater economic vulnerability than was experienced in the north and west. Unfortunately, there is little or no evidence concerning the extent to which fertility control was an option for rural couples in mid-nineteenth century England.

The question of access to the means of fertility control becomes even more crucial if one considers other possible explanations for the marital fertility differential. Probably the simplest of these appeals to the fact that age-specific fertility within marriage is associated with marriage duration (and hence with age at marriage). At any given age, couples who have been married for a shorter time have higher fertility (Knodel, 1978: pp. 492–502; 1988: p. 372–87). Thus the higher marital fertility of couples in the north and west of England and Wales may simply be the result of their later age at marriage. In the absence of deliberate birth control, the relationship between marital fertility and marriage duration has normally been ascribed to greater frequency of intercourse or a lower prevalence of secondary sterility amongst couples married for a shorter time (although there is little or no evidence for either of these mechanisms).

Whilst this second explanation sounds plausible, and is seemingly backed up by hard evidence from the period prior to 1800 (Wrigley & Schofield, 1983: Tables 9 and 10, pp. 173 and 174), it rests upon the assumption that deliberate birth control, or 'stopping' behaviour does not take place. Whilst this assumption may be reasonable for the seventeenth and eighteenth centuries, it becomes more risky as the nineteenth century progresses, and the era during which we know 'stopping' behaviour to have been usual becomes closer. In order to be sure of its reasonableness or otherwise in 1861, we again need information on the availability of methods of fertility control at that time.

Between 1861 and 1911, the evidence we have suggests that there was a parallel decline of marital fertility in the two zones of England and Wales. It seems, therefore, that although the level of marital fertility may have exhibited geographical differentials, its temporal trend did not. Those holding the 'demographic monolith' view of England and Wales during the fertility transition would appear to be partially vindicated. But this appearance may be false. Recall that at the beginning of this paper we stressed two aspects of the fertility transition: the decline in the level of fertility within marriage, and the change in fertility behaviour that accompanied it (the adoption of voluntary fertility control). The critical point is that even though the timing of the decline in marital fertility in the two areas is the same, this need not have been true of the onset of voluntary fertility control especially if the absolute levels of fertility differed between the two areas during the period of decline. In the simplest case, if we assume that the onset of such behaviour is associated with the attainment of a certain family size, then clearly this will be attained at different times in two such areas. In reality, such a simple assumption is not justified, and the relationship between the level of marital fertility and the extent of voluntary fertility control within marriage is rather complex (see Knodel, 1988: pp. 287–317 for a case study and a discussion). Despite this, it is incorrect to conclude that contemporaneity in the timing of the fertility decline in two areas implies contemporaneity in the timing of the onset of voluntary fertility control.

It might be asked whether or not there is any direct evidence of such differentials in the timing of the onset of voluntary fertility control. In previous work (Hinde, 1985a: pp. 272–306) I attempted to measure the extent of fertility control within marriage in two small communities, one in Norfolk and one in Shropshire. The study only covered the period up to 1881, and the results should be interpreted very cautiously since they are based on small numbers of married couples. With these reservations in mind, the analysis revealed *some* evidence of fertility control in Norfolk during the late 1870s amongst couples in which the husband was an agricultural day labourer. Amongst similar couples in Shropshire, there was no such evidence. We might speculate, therefore, that the greater tendency of households in the south and east of England to be subject to periodic economic 'crises' led to couples there adopting 'stopping' behaviour earlier than couples in the rest of England and Wales (see Hinde, 1985a: pp. 307–72 for a fuller exposition of this idea).

Conclusions

The conclusions of this paper are threefold. First, there were significant geographical differentials in the levels of both nuptiality and marital fertility in rural England and Wales during the period of the fertility transition between 1871 and 1931. Second, the regional pattern of nuptiality was associated with that of the structure of the agricultural work force. Third, under the household formation system prevailing in nineteenth century England and Wales, agricultural work patterns may have influenced fertility behaviour through their effect on household resources. In order to examine the last of these speculations further, we require further analyses of fertility (preferably at the individual level) for the second half of the nineteenth century.

Acknowledgements

I am indebted to Professor R. I. Woods of the Department of Geography at the University of Liverpool for his generosity in making available some of the data used in this paper. I should also like to thank the participants at the symposium for their comments. Any errors or inaccuracies, of course, are my responsibility.

References

Anderson, M. (1976). Marriage patterns in Victorian Britain: an analysis based on registration district data for England and Wales 1861. *Journal of Family History*, **1**, 55–78.

Brass, W. & Kabir, M. (1977). Regional variations in fertility and child mortality during the demographic transition in England and Wales. In *Regional Demographic Development*, ed. J. Hobcraft and P. Rees. pp. 71–88. London: Croom Helm.

Census of England and Wales (1871). *Population Tables – Area, Houses and Inhabitants, III. Population Abstracts; Ages, Civil Conditions, Occupations, and Birth-Places of the People.* (British Parliamentary Papers, 1871/LXXII) [Cd. 872].

Coale A. J. (1967). Factors associated with the development of low fertility: an historic summary. In *Proceedings of the World Population Conference, 1965, Volume 2*, 205–9. New York: United Nations Department of Economic and Social Affairs.

Davis, K. (1963). The theory of change and response in demographic history. *Population Index*, **39**, 345–66.

Friedlander, D. (1973). Demographic patterns and socioeconomic characteristics of the coal mining population of England and Wales in the nineteenth century. *Economic Development and Cultural Change*, **22**, 39–51.

Friedlander, D. (1983). Demographic responses and socioeconomic structure: population processes in England and Wales in the nineteenth century. *Demography*, **20**, 249–72.

Haines, M. R. (1979). *Fertility and Occupation: Population Patterns in Industrialization*, London: Academic Press.

Hajnal, J. (1982). Two kinds of preindustrial household formation system. *Population and Development Review*, **8**, 449–94 (Reprinted in *Family Forms in Historic Europe*, (1983). ed. R. Wall, J. Robin and P. Laslett, pp. 65–104. Cambridge: Cambridge University Press).

Hinde, P. R. A. (1985a). The fertility transition in rural England. Unpublished Ph. D. thesis, Department of Geography, University of Sheffield.

Hinde, P. R. A. (1985b). Household structure, marriage and the institution of service in nineteenth-century rural England. *Local Population Studies*, **35**, 43–51.

Hinde, P. R. A. & Garrett, E. M. (in press). Work patterns, marriage and fertility in late-nineteenth centurty England. In *Regional and Spatial Demographic Patterns in the Past*, ed. R. M. Smith. Oxford: Basil Blackwell.

Knodel, J. E. (1978). Natural fertility in pre-industrial Germany. *Population Studies*, **32**, 481–510.

Knodel, J. E. (1988). *Demographic Behavior in the Past: a Study of Fourteen German Village Populations in the Eighteenth and Nineteenth Centuries.* Cambridge: Cambridge University Press.

Kussmaul, A. (1981). *Servants in Husbandry in Early Modern England.* Cambridge: Cambridge University Press.

Snell, K. D. M. (1985). *Annals of the Labouring Poor: Social Change and Agrarian England, 1660–1900*. Cambridge: Cambridge University Press.

Teitelbaum, M. S. (1984). *The British Fertility Decline: Demographic Transition in the Crucible of the Industrial Revolution.* Princeton: Princeton University Press.

Woods, R. (1982). *Theoretical Population Geography.* London: Longman.

Woods, R. (1986). The spatial dynamics of the demographic transition in the West. In *Population Structures and Models: Developments in Spatial Demography*, ed. R. Woods & P. Rees, pp. 21–44. London: Allen and Unwin.

Wrigley, E. A. & Schofield, R. S. (1983). English population history from family reconstitution: summary results 1600–1799. *Population Studies*, **37**, 157–84.

8 *Fertility decline and birth spacing among London Quakers*

J. LANDERS

The decline in the marital fertility levels of European populations from the later nineteenth century[1] has long been recognized as a landmark in the demographic evolution of the continent, and has been made the subject of a correspondingly detailed scrutiny[2]. Particular interest has attached to the behaviour of the so-called 'forerunners', groups which acted as a kind of demographic vanguard reducing their fertility substantially before that of the population at large (Livi-Bacci, 1986). Research on such groups has revealed that they were predominantly urban and that they were usually 'bounded' in some way, that is to say they were clearly distinguished from the surrounding population, whether by their enjoyment of a shared socio-legal privilege or by their membership of a minority religion.

In this context the experience of London's Quakers is of particular interest, for in 1861, on the eve of the national fertility decline, the registration districts with the lowest marital fertility were all to be found in the capital (Woods & Smith, 1983) whilst in North America Quakers were apparently one of the first groups to reduce their fertility substantially. In a study of 276 reconstituted families from the middle colonies, Wells (1971) found signs of this among women born 1731–55 and it had become general among the cohort born 1756–85. A recent study by Byers (1982) has also demonstrated the existence of family limitation in Quaker dominated Nantucket from the middle of the eighteenth century.

The present study is based on a family reconstitution of two of London's six 'Monthly Meetings' of Quakers[3], from which it was possible to obtain age-specific fertility and mortality rates, together with data on age at marriage. The latter reveal that Quakers marrying in London were generally older than the national average for the population at large, although after 1700 the brides were somewhat younger than seems to have been the case among Quakers outside the capital (see Table 8.1). There were generally too few recorded marriages of London born women in our study to merit separate analysis, but in the one case where this was possible, with the cohort marrying 1700–49, the mean age was 22.0 years ($n=49$), a similar figure to that found in studies of seventeenth century London (see Table 8.2).

Table 8.1(a). *Male and female ages at marriage*

	London Quakers		*Southern English Quakers*	
	Males[a]	*Female*[b]	*Males*[a]	Females[b]
1650–99	31.0	27.1	28.6	24.9
1700–49	29.4	25.6	28.1	26.3
1750–99	29.5	26.5	30.0	27.5
1800–49	30.1	26.5	31.7	29.8

[a] Excluding known re-marriages, [b] First marriages only.
Source: Reconstitution tabulations and Eversley (1981: p. 65).

Table 8.1(b). *Means of 12 English Reconstitutions (first marriages)*

	Males	*Females*
1650–99	27.8	26.5
1700–49	27.5	26.2
1750–99	26.4	24.9
1800–49	25.3	23.4

Source: Wrigley & Schofield (1981: p. 255). For a revised age at marriage series, using 25 year cohorts and incorporating the results of aggregative back projection, see Schofield (1985). The trend in the series is not, however, affected by the new estimates.

Table 8.2. *Female age at first marriage in London*

	Age	*N*
Parishes		
(London born women only)		
St. Peter's Cornhill, 1580–1650	23.7	15
St. Michael's Cornhill, 1580–1650	21.3	22
St. Mary Somerset, 1605–53	24.7	9
St. Botolph Bishopsgate, 1600–50	23.8	24
Diocese of London, 1598–1619		
London born	20.5	496
Migrants	24.2	500

Source: Elliott (1981: p. 87).

There was little change in mean age at marriage over the period, despite the marked reduction visible in the national series, leading by the early nineteenth century to a considerable gap between the London Quaker figure and that found in other English reconstitutions. Fertility within marriage, however, fell markedly in the later eighteenth century and showed only a limited recovery after 1800 (see Table 8.3). This reduction, absent from other English data (see Figure 8.1), raises the question of conscious fertility control and the possible 'forerunner' role of our study population. It has been suggested that the Quaker 'discipline' may have been particularly favourable to the development of low fertility within marriage. Theologically, Friends differed from other groups in their attitude to the status of young children. As Byers (1982: p. 40) argues:

> Quaker ideology may have strengthened existing desires for fewer children through its assumption of childhood innocence. Quakers believed that neither sin nor grace were conveyed by heredity but that upon coming of age a child becomes susceptible to sin and depravity. This belief in childhood innocence implies a certain amount of maleability – that a child can be taught to remain open to God's grace and to resist the world's evil.

The stress on 'nurture' led, it is suggested, to a desire for smaller families so that parents could concentrate more attention on each child. There is also some evidence that the discipline 'fostered sexual prudery within marriage and resulted in a decline in the amount of intercourse between husbands and wives' (Frost, 1973: p. 70).

Quakers' readier acceptance of family limitation may be explained in these terms, but they cannot in themselves account for the timing of this development or for differences between the fertility levels of Quaker communities in different regions. To do this we need to consider the same social, economic and demographic factors as are invoked in fertility analysis of the wider society, but the character of the discipline cannot be discounted in such an analysis, for the role of economic circumstances in shaping Quaker family life was itself mediated by that of the monthly meeting. As Frost (*ibid*: p. 65) states:

> If the immediate family became financially insolvent, the meeting might give money, make loans, seeds tools or perhaps a cow so that the family might survive. If there were too many brothers or sisters for the parents to support the meeting could take on the responsibility of placing some of the children in other good Quaker homes.

In an urban community such material support would generally consist of raw materials or stock in trade, but the reference to 'too many brothers and sisters' is thought provoking. Monthly meetings made extensive demands on members in the ordering of their affairs (see Landers, 1984: Chapter 3).

Table 8.3. *London Quakers: age-specific marital fertility rates per thousand (number of years observed in brackets)*

Age group	*Cohorts* 1650–99	1700–49	1750–99	1800–49
20–4	250 (12)	412 (131)	403 (62)	463 (54)
25–9	450 (20)	429 (170)	319 (91)	260 (100)
30–4	300 (30)	329 (149)	221 (104)	267 (90)
35–9	263 (19)	197 (132)	113 (97)	254 (67)
40–4	133 (15)	69 (116)	94 (96)	43 (47)

Source: Reconstitution tabulations.

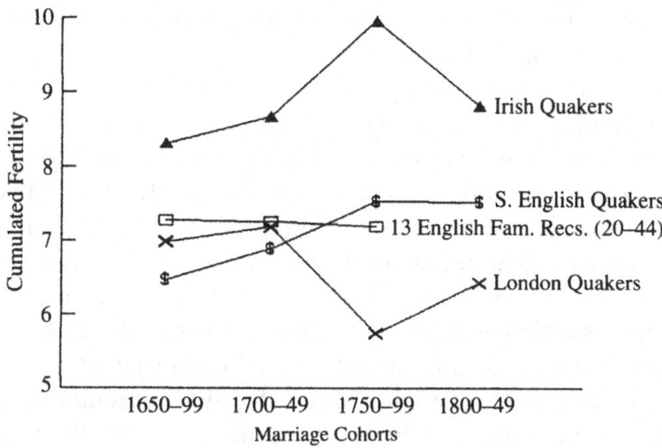

Figure 8.1. Cumulated age-specific fertility rates, ages 20–39 years.

In the field of family formation these centred on the regulation of marriage, but the Quakers' 'communal risk sharing' (Smith, 1981) may have affected fertility as well as nuptiality. We shall return to our family reconstitution results with this possibility in mind, but first we must look at the theoretical framework which demographers have used to analyse the process of fertility decline and particularly at the ambiguous relationship between concepts of fertility regulation in pre- and post-transitional populations.

Natural fertility and family limitation

Demographers have often approached the historical process of fertility decline (the so-called 'fertility transition') in terms of concepts formulated by Louis Henry, attributing it to an underlying qualitative change in patterns of family formation. The original aim of Henry's pioneering work was to obtain a 'datum line' for uncontrolled human fertility, against which levels recorded in contemporary populations could be compared. In order to obtain the necessary information he turned to historical sources and developed the method of family reconstitution so as to utilize them fully[4].

The resulting estimates of marital fertility in pre-transitional populations proved, however, surprisingly variable and Henry was forced to abandon the notion of a fixed 'physiological' fertility pattern in favour of what he termed 'natural fertility'[5]. Under 'natural' conditions fertility might vary appreciably, in response to biological or social factors, but such variation was seen as qualitatively different from that characterizing the fertility of post-transitional populations. The latter, he argued, arose from the practice of what was termed 'family limitation' according to which couples altered their reproductive behaviour with changing parity, as they approached a particular desired family size.

Natural fertility thus refers to fertility in the absence of parity-specific regulation, a definition by exclusion which might lead one to expect that it would denote simply a residual category, encompassing different fertility regimes with varying levels of conscious control. But Henry and many subsequent writers have rejected this interpretation, arguing that natural fertility regimes can be identified on positive criteria and that the conscious control of fertility can be wholly equated to the practice of family limitation. This assertion was based on two grounds, the first being technical, whilst the second had a broader sociological character.

From the technical point of view it was noted that schedules of age-specific fertility rates, obtained in the absence of family limitation, exhibited a characteristic convexity when plotted against maternal age. This convexity, which was wholly independent of the absolute level of fertility, was quite distinct from the age pattern of fertility in populations practising parity-specific control, where a marked concavity develops.

On a larger scale, Henry linked the adoption of family limitation to the transition from the 'traditional' to the 'modern' social worlds, implying that its appearance marked the entry of rationality into the sphere of reproductive behaviour[6]. Implicit in this view, but explicit in those of later writers such as John Knodel (1977), was the claim that deliberate long term control of fertility within marriage was itself a product of 'modernization'. Hence the two forms of fertility regime could be seen as the embodiment of two opposing states of society and social mentality, in one of which reproduction was the outcome of conscious calculation and instrumental rationality, whilst in the other it was a matter of unreflecting cultural automatism[7]. In Knodel's words (Knodel, 1988: p. 455):

> It is this shift from natural fertility to family limitation that constitutes a fundamental break with past patterns of marital childbearing and represents the modernisation of reproductive behaviour.

Subsequent research on this question has, however, brought mixed results. Whilst the hypothesis of a characteristic age pattern of natural fertility has been strongly validated, the radical distinction between 'traditional' and 'modern' fertility regimes has remained the subject of disputes[8]. The most powerful evidence on the former point has come from the successful construction of theoretical models able to describe a wide range of observed fertility schedules in terms of a small number of parameters. Coale and Trussell (1974, 1978) obtained remarkably good results with a two parameter model which has since furnished demographers with a rigorous quantitative technqiue for the detection of family limitation among populations for whom only statistical evidence is available.

Knodel made use of this technique to demonstrate the absence of parity-specific family limitation from a number of pre-transitional populations, both historical and contemporary, for which age-specific marital fertility rates could be obtained. On this basis he put forward a strong version of the 'traditional fertility' thesis, arguing that family limitation was not only unknown in such societies, but that any deliberate long term regulation of fertility within marriage was simply unthinkable (Knodel & Van de Walle, 1979: p. 229).

The issues involved here may be illuminated by a simple diagrammatic model (see Table 8.4). We can divide those forms of behaviour which serve to reduce marital fertility into two categories: those which do so by terminating childbirth prematurely ('stopping') and those which operate by prolonging birth intervals without reference to parity ('spacing'). In the same way, we can label fertility regimes as 'rational' or 'traditional' according to whether fertility, in Coale's well known phrase, lies inside or outside 'the calculus of conscious choice' (Coale, 1973).

This results in four permutations (see Table 8.4), the first of which, type I, corresponds to Henry's parity-specific family limitation whilst type II is found, for example, where cultural rules proscribe continued childbearing by grandmothers. Knodel's argument turns on the categorization of fertility regimes based on 'spacing'. If we accept that parity-specific control is absent from pre-transitional populations then the 'traditional fertility' thesis requires that all parity-independent restrictions should be allocated to type IV, arising from cultural or religious prescriptions, periods of ritual abstinence, direct ecological or dietary constraints etc., since type III regimes are defined out of existence.

Table 8.4. *Strategies of fertility control*

(for explanation see text.)

	Mode of fertility control	
	'Stopping'	*'Spacing'*
Mode of social action		
'Rational'	I	III
'Traditional'	II	IV

This necessity arises from the marked regional variability of marital fertility in pre-transitional populations (see Coale & Watkins, 1986: Map 2.2), the significance of which was first stressed by Carlsson (1966) using Swedish data. If such variability is a reflection of underlying differences in conscious strategies of family formation then our understanding of the fertility transition itself must be greatly altered. From a qualitative restructuring of attitudes and behaviour this process becomes a merely quantitative reorientation to changing circumstances: in Carlsson's own words, 'innovation' gives way to 'adjustment process'.

Knodel cites three reasons for maintaining the 'sociological' distinction between parity-specific and parity-independent effects. 'Spacing' behaviour, he argues, is 'not always clearly volitional' (Knodel, 1977: p. 241) and even where this is the case it 'may result from a *reaction* to immediate or recent adversity rather than from an anticipation of long-range benefits such as characterise family limitation' (*ibid*). Finally, since parity-independent regulation is 'typically temporary' the technique employed need only be acceptable for a short period. This:

> could be important, because couples might be willing to resort to abstinence, the most universally obvious and accessible method of birth control, in times of crisis or for short periods of time even when unwilling to forgo sexual relations permanently for limiting purposes.

(*ibid.*)

According to this view the early stages of the fertility transition, at least, are dominated by the innovative force of family limitation as a newly perceived or at least newly acceptable 'behavioural norm' (Knodel, 1977: p. 249), which is associated with fertility becoming 'a matter for rational choice' (Knodel, 1977: p. 247).

Reasoning of this kind, according to which pre- and post-transitional fertility regimes are qualitatively distinct, and the passage between the two driven, or at least mediated, by 'ideational' change, is challenged by those who prefer to stress the role of changing economic circumstances in triggering fertility decline. One of the most influential partisans of this latter view has been J. C. Caldwell who bluntly asserts that fertility in all societies is economically rational, claiming that the basis for 'high fertility' among pre-transitional populations is to be sought in a family structure which channels benefits from children to parents (Caldwell, 1976; 1982). Caldwell sets out to overturn the concept of the 'traditional' high fertility regime, maintained by a series of normative 'props' supporting behaviour which runs counter to the economic interests of the parties concerned, and although his 'wealth flows' theory has attracted criticism this has dwelt as much on the specific benefits accruing to high fertility as on the notion that such fertility is a 'rational' response to the structural context (Cain, 1982).

The advocacy of such a view in a Third World context results, however, in something of a paradox, for marital fertility in many of these societies is appreciably lower than was the case in those parts of pre-transitional Europe where the nuclear family household system reduced the economic benefits of parenthood substantially[9]. This 'low' fertility arises of course from the extension of birth intervals due to prolonged lactation and postpartum abstinence. The Caldwells (Caldwell & Caldwell, 1981) review a number of explanations given by their Yoruba informants for the latter practice. These include a concern for the welfare of the infant, together with a number of 'moral', 'magical' or religious considerations, but the authors themselves stress the important sociological function of sexual abstinence in protecting the 'patriarchal' lineage from the subversive effects of close sexual and affective ties between spouses on their loyalties to wider kin.

Although the theory of wealth flows might, at least in principle, be able to explain early marriage, or the non-acceptance of Western style birth control, it has ambiguous implications for our understanding of pre-transitional marital fertility regulation and for the conceptual relationship between such regulation and that seen in post-transitional populations. On the one hand Caldwell argues that the dictates of economic rationality in, for instance, Yoruba society prescribe very high levels of fertility, but these are not achieved in practice because of the constraints imposed by the kinship system. Hence the fertility regime in these circumstances cannot be identified unequivocally with either term of the 'grand dichotomy'.

The rejection of family limitation can thus be explained in terms of an economic rationality quite incompatible with the 'traditional' category, but at the same time the play of such rationality is subordinated to the demands of extended kinship solidarity in a manner wholly alien to the supposed characteristics of 'modern' behaviour, according to which individuals, or couples, pursue their self-interest without reference to such considerations[10]. Hence Caldwell's arguments could equally well be used to place the Yoruba regime, and others like it, in category III or IV of our diagrammatic model.

The detailed studies of European fertility change carried out under the auspices of the Princeton Office of Population Research have also thrown some doubt on the notion that sustained decline in fertility must necessarily represent a sharp break with the attitudes and behaviour of former times. Demeny (1972), for instance, found that data from the provinces of Austria–Hungary in the 1880s yielded values of I_g, Coale's index of marital fertility, which varied between 0.46 and 0.9.

Variations of this magnitude may, as Knodel insists, be compatible with the absence of parity-specific control, but such a conclusion simply raises once more the question of how we should see the relationship between this type of fertility regulation and that which is exercised by 'spacing'. Livi-Bacci (1977: p. 285), discussing the moderate levels of marital fertility in much of nineteenth century Italy, concluded:

> that concern for the dimensions of the family was not alien to the psychology of the Italian couple prior to the beginning of the decline, and that their natural fertility was influenced by a conscious effort to prolong inter-birth intervals.

The existence of markedly low fertility among certain restricted groups, he argued, demonstrated both that 'groups sufficiently motivated had the knowledge and the necessary will to lower fertility' and that the 'roots of behaviour modification often come into being far before a particular change becomes apparent in any way' (*ibid*). Whether these 'roots' lay in that 'concern for the dimensions of the family' identified in the wider society, or were a particular characteristic of a modernizing minority, remained, however, an unresolved question.

The fertility of London Quakers

These theoretical questions have important implications for any discussion of the empirical results obtained from our family reconstitution study. The efforts of mathematical demography have resulted in sophisticated techniques for the analysis of structural distinctions between patterns of fertility, but the broader significance of these distinctions is far from clear. We cannot therefore assume, as might once have been possible, that the absence of family limitation before a given date necessarily indicates a lack of conscious control over fertility, nor can we reduce the general problem of fertility change to that of a passage from natural fertility to family limitation.

The practice of family limitation is, nontheless, the first possibility which needs to be investigated if we are to explain the reduction in fertility of the reconstituted families after 1750. It can be detected using the Coale–Trussel model based on the expression:

$$r(a)/\mathrm{n}(a) = M \exp m \cdot v(a)$$

where $r(a)$ is the observed rate of marital fertility, at age a, in a given population, and $n(a)$ is the equivalent rate from a model fertility schedule. The quantities $v(a)$ refer to a schedule of age-specific deviations from the natural fertility curve, calculated by Coale and Trussel (1974) using a set of fertility schedules obtained from populations practising family limitation.

The model yields values for the two parameters M and m, of which m, indicates the degree of parity-specific fertility reduction present in the observed population, whilst M is a scale factor related to the underlying level of natural fertility, or 'fecundity' in the sense of Knodel & Wilson (1981). The interpretation of any given value of m is not entirely straightforward, but the authors suggest that values below 0.2 indicate the absence of family limitation, whilst a value of unity indicates a prevalence equal to that among the populations used to construct the schedule of $v(s)s$.

The model can be fitted to an observed schedule using the formula:

$$\mathrm{Ln}(r(a)/n(a)) = \mathrm{Ln}M + mv(a)$$

The mean square error (MSE) provides a measure of goodness of fit, although here again any given value of the MSE lacks a rigorous interpretation and Coale & Trussel suggest simply that a value of around 0.005 'indicates a mediocre fit and a value of 0.01 indicates a terrible fit' (Coale & Trussel, 1978: p. 204). The procedure for fitting the model is to assume that $v(a)$ is zero for the 20–24 year age group, in which $n(a)$ reaches a maximum. This presents a difficulty in the present case, for the maximum observed fertility rate, prior to 1750, occurs not in this age group but in the subsequent one, ages 25–29 years.

This is a feature of a number of fertility schedules from English reconstitutions in the period before 1750 (see Flinn, 1981: pp. 102–9) and Wrigley (1978a) suggests putting $v(25–29)$ equal to zero and then taking $r(20–24)$ from the model schedule proportionate to $r(25–29)/n(25–39)$ in such cases. This procedure was followed to obtain the results in Table 8.5. The fit, as judged by the MSE, is extremely poor but much better results were obtained by omitting the fertility rate in the 40–44 year age group. The model now suggests an appreciable level of family limitation in the first half of the eighteenth century (the seventeenth century rates are based on too few observations to justify analysis), rising sharply after 1750, with a base level of fecundity close to the average of the schedules used to construct the $n(a)$ series.

Table 8.5. *Coale–Trussel marital fertility parameters*

London Quakers (all marriage ranks). (For explanation see text.)

	Fertility				
	ages 20–44 years:			*ages 20–39 years*	
	*1700–49**	*1750–99*	*1800–49*	*1700–49**	*1750–99*
m	0.63	0.45	0.68	0.48	0.87
M	1.12	0.80	0.97	1.07	0.92
MSE	0.009	0.004	0.095	0.004	0.004

*20–24 year old fertility rate obtained from model schedule.

The fall in the total fertility rate might thus be explicable in terms of the practice of parity-specific control, but the fit obtained between data and model, even by eliminating the rates for the oldest age group, is little better than 'mediocre', and this elimination results in parameter estimates for the 1700–49 cohort which are based on only two data points. And in addition to these technical considerations there is also some internal evidence which suggests that the changing patterns of family formation were more complex than those implied by the Henry–Knodel model of fertility transition.

We can approach this question by turning our attention from age-specific fertility rates to another fertility measure[11], completed family size (see Figure 8.2). These results indicate a marked fall in fertility from the first to the second century of the study, with much less change within either of the two sub-periods. An analysis of famiiy size broken down by age at marriage, however, reveals some striking divergences in the behaviour of the different bridal age groups. Thus the family sizes of the younger brides decline rather more after 1700 than the aggregate figures would suggest, but the most dramatic feature of the table is the near halving of family sizes among women marrying at ages above 25 years in the later part of the eighteenth century, a much greater reduction than is displayed by women marrying below this age.

This phenomenon can be investigated through an analysis of birth intervals. Birth interval analysis is a very powerful technique in this context since it is possible to make use of a much larger proportion of the original data without risk of biasing the calculations and the method is very sensitive to variations in the structure of marital fertility. The practice of family limitation, as described by Henry, has some very specific implications for the distribution of birth intervals. The stipulation that couples

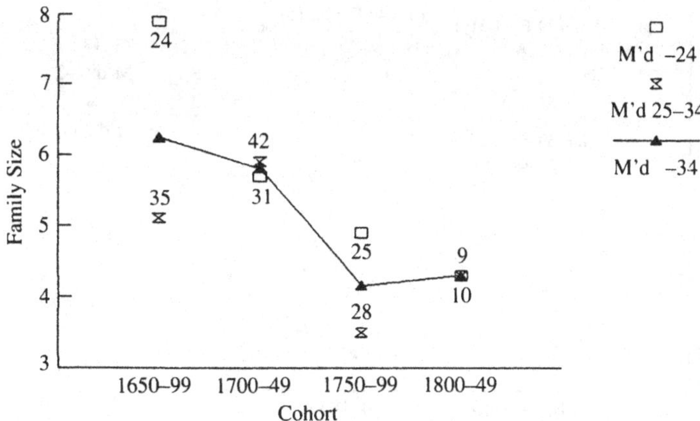

Figure 8.2. Completed family size by marriage age (with numbers observed).

vary their reproductive behaviour as they approach their desired family size implies a marked reduction in the 'tempo' of fertility with increasing durations of marriage and thus the lengthening of higher order birth intervals relative to those of lower order.

The spread of family limitation may, in principle, make little difference to low parity birth intervals but it has often been argued that those of the highest parities will become both longer and more variable, reflecting accidental conceptions, 'second thoughts', or deliberate conceptions intended to 'replace' a child who has recently died (Wrigley, 1966b: pp. 93–4). The 'spacing' model of fertility control, by contrast, implies a more even reduction in the tempo of fertility and a lengthening of the lower parity birth intervals.

A useful measure of the tempo of fertility in the earliest stages of family formation is provided by the estimation of fecundability at marriage. Fecundability, a concept introduced by Gini, is defined as the probability of conception per cycle of a woman, cohabiting with a man, and not attempting to prevent conception in any way (Henry, 1976: pp. 230–47). It can be estimated from the distribution of protogenesic intervals using a method devised by Bongaarts and modified somewhat by Knodel and Wilson (1981). In this method the proportion of intervals of fewer than eleven months, excluding those of less than eight months, is obtained and used as the point of entry to a table of fecundabilities constructed by the authors from a theoretical distribution function.

The intervals obtained from the reconstruction were pooled into two cohorts, so as to obtain larger numbers for the analysis, and then split into bridal age groups 25–34 years and under 25 years so as to obtain the results

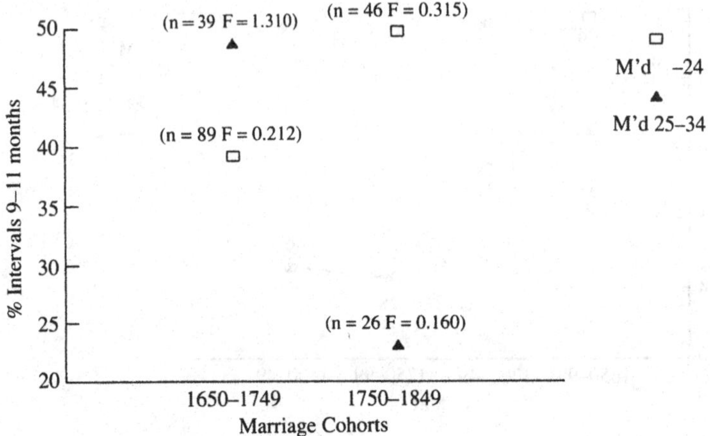

Figure 8.3. Percentage of 'short' protogenesic intervals by age of marriage.

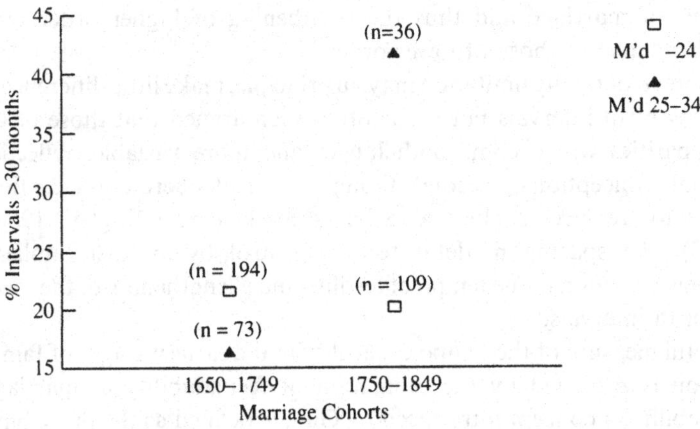

Figure 8.4. Percentage of 'long' birth intervals by age of marriage: parities < 4.

in Figure 8.3. The results for the initial cohort indicate a rather higher level of fecundability among the older brides and a markedly lower level among those marrying at ages below 25. After 1750 this picture is reversed. The fecundability of the latter group rises sharply, whilst that of the 25–34 groups falls below the lower limit of the Knodel–Wilson tabulation[12]. High fecundability early in marriage is compatible with parity-specific 'stopping'

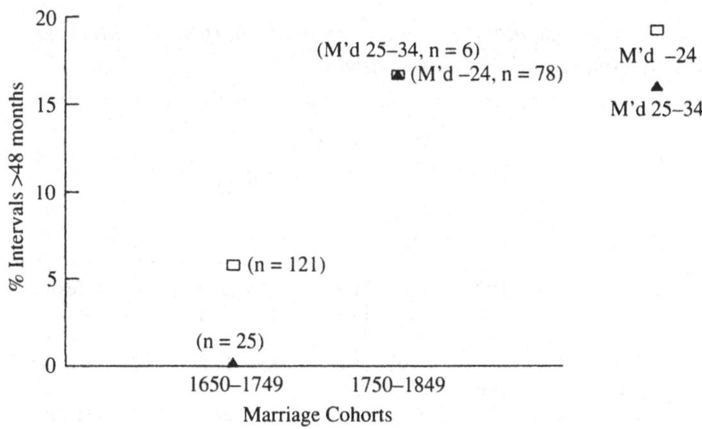

Figure 8.5. Percentage of 'long' birth intervals by age of marriage: parities 4+.

strategies of family formation, but the behaviour of the older women is hard to explain in such terms particularly because family limitation and delayed marriage are so often seen as functional alternatives to each other[13].

The distributions of intergenesic intervals reveal a similar pattern. It has been suggested by Dupâquier (Dupâquier & Lachiver, 1969) that increases in the proportion of intervals of more than 30 months, and especially of those longer than 48 months, signify the adoption of family limitation in a population, although such a criterion does not take account of parity. Hence in Figures 8.4 and 8.5 we plot the frequency of 'long' intervals at high or low parities, taking 48 months and 30 months respectively as our criterion.

The data for the first cohort confirm the faster tempo of fertility among women marrying over 25, but fewer than six percent of high order birth intervals among the younger brides fall into this category. This proportion nearly triples after 1750, but the frequency of 'long' low order birth interval among the older age group reflects the decline in fecundability and is now significantly greater than that observed in the under 25 group[14].

The average intervals tabulated by 50 year cohorts (see Table 8.6) are roughly consistent with these findings, though based on smaller numbers, but they suggest a revival in the tempo of fertility among older brides in the nineteenth century.

Table 8.6. *London Quakers: mean and median intergenesic intervals (months) by cohort and age at marriages*

	Age at Marriage								
	–24				*25–34*				
Parities:	*1–3*	*(n)*	*4+*	*(n)*	*1–3*	*(n)*	*4+*	*(n)*	
1650–99	23.5		31.3		21.9		26.0		Mean
	18.1	(42)	32.0	(16)	13.8	(15)	27.0	(3)	Median
1700–49	25.0		24.2		21.0		20.2		Mean
	21.3	(152)	22.1	(105)	18.0	(58)	17.0	(22)	Median
1750–99	24.6		35.6		25.4		22.5		Mean
	20.4	(71)	24.5	(43)	30.7	(1)	24.0	(2)	Median
1800–49	27.3		30.1		25.4		36.3		Mean
	26.0	(38)	23.2	(35)	21.0	(18)	27.0	(4)	Median

Strategies of family formation

The above analysis supports the existence of family limitation among women marrying below the age of 25, but the results are more consistent with a 'spacing' strategy where the older brides are concerned, at least in the 1750–99 cohort. This age group may have adopted parity-specific control after 1800, but the nineteenth century data are rather scanty. The picture before 1750 is also ambiguous. The Coale–Trussell model suggests a moderate degree of family limitation in the 1700–49 cohort and women marrying at the younger ages in these decades had smaller families than had been the case in the seventeenth century. The birth interval analysis, however, suggests that the spacing model is more appropriate here and the position is further complicated by the finding of relatively low age-specific marital fertility among younger women elsewhere in England at this time.

Two points, however, stand out clearly and require some explanation. These are; firstly, the sharp reduction in marital fertility which is visible among the cohorts marrying after 1750 and secondly, the markedly different ways in which this reduction was accomplished by women in the different marital age-groups. In particular we have to account for the adoption of parity-specific control by the older wives and thus confront the theoretical problems we reviewed in an earlier section. The age-specific character of the spacing behaviour, together with its rapid appearance from one cohort to the next, leaves little room for doubt that it was volitional in nature. In the same way, it was evidently sustained over a term of years and can hardly be dismissed as a short term reaction to events.

The central reason advanced by Knodel for seeing parity-specific control as *the* rational strategy of family formation is that first formulating, and then adhering to, a 'target' family size involves assessing costs and benefits over the long term in a way that merely delaying births does not (Knodel, 1977: p. 241–2). What is not clear, however, is that long term 'accounting' of this kind exhausts the possibilities for economically rational fertility behaviour. Instead, we argue that the temporal and other criteria employed in 'rational' decision making are context dependent and that there are circumstances under which completed family size may be of little relevance to this process.

Discussions of fertility control have tended to resolve around the question of completed family size, partly because this is such an 'obvious' demographic variable, and partly because theoretical treatments of the problem have dwelt particularly on questions of inheritance and 'heirship strategies' (Wrigley, 1978b). Recent research on early modern England, however, has led to a greatly reduced emphasis on the demographic significance of inheritance, whilst enhancing that of a normative living standard or, 'culturally determined moral economy' (Schofield, 1989; Smith, 1981). The Quaker discipline emphasized this factor pre-eminently whilst reducing the significance of inter-generational wealth flows once the children had left home.

If we assume that children were economically 'neutral' once they ceased to be entirely dependent then the numbers surviving to adulthood, or to the death of the parents, is relatively insignificant. Instead we can construct a model based on the numbers of children who are dependent on their parents and present in the the parental household at any one time. A rational fertility strategy, on this criterion, is one which minimizes the maximum number of dependents, or alternatively which minimizes the period during which large numbers are present.

A strategy of this kind would seem more in keeping with the pre-occupations of the monthly meeting and with the economic circumstances of a population of small traders and artisans, than would be one based on final parity. Its implication can be mapped out in very broad terms, with the aid of some simple calculations. If we assume a base level of fecundity yielding protogenesic birth intervals of 12 months and intergenesic intervals of 24 months, with an effective child bearing period stretching into the late 30s, then we can determine the consequences of a given reduction in total fertility achieved either by 'spacing' or 'stopping', in the absence of mortality.

Where children cease to be dependent at ten years of age (see Figure 8.6) then the extension of birth intervals by 50 per cent is clearly more effective in lowering the number of years during which large numbers are present than

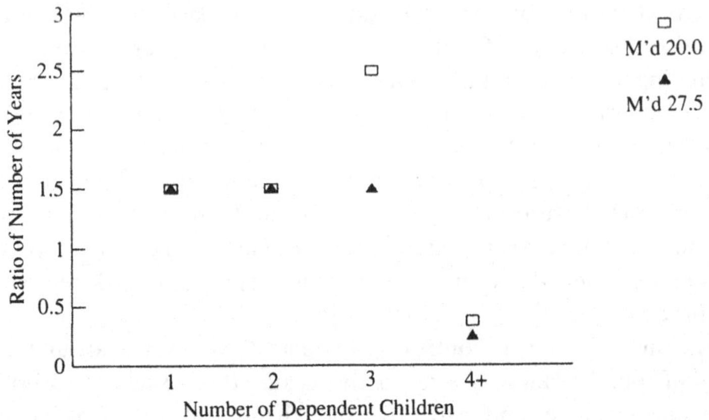

Figure 8.6. Child years of dependency: ratios of 'spacing' to 'stopping': dependency to age 10.

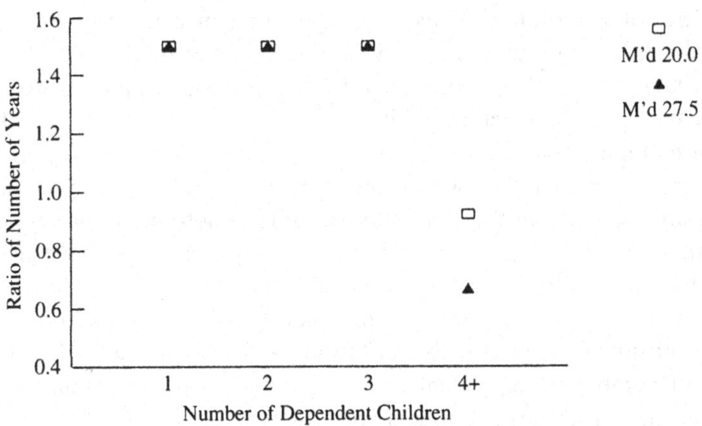

Figure 8.7. Child years of dependency: ratios of 'spacing' to 'stopping': dependency to age 15.

is a strategy based on the same final parity achieved at the 'natural' tempo. The period in which more than three children are present is reduced by between 60 and 70% depending on age at marriage. The extension of the period of dependency to 15 years (Figure 8.7), however, changes the picture considerably. 'Spacing' remains more efficient than 'stopping' for women who marry in their later twenties, but it loses its advantages for those who marry earlier and it imposes a lengthier period of 'intermediate' dependency.

The model we have outlined, although highly simplistic, produces exactly the divergence which we have observed among the 1750–99 marriage cohort, given a finite but rather long period of dependency. A detailed investigation of childhood and adolescence among the London Quakers lies outside the scope for the present study, but we do have some evidence to suggest that children did remain in their parents' household for extended periods in the central decades of the eighteenth century.

This evidence comes from the membership lists which Southwark Monthly Meeting drew up in 1737 and again in 1762. These lists are arranged in household 'blocks' with children, servants and apprentices identified as such. These household entries were linked to the family reconstitution forms so as to determine the proportions of surviving children entered in the lists at various ages. The results suggest some extension of the period of dependency between the two dates, but even in 1737 more than 75% of surviving teenage children are listed as members of their parents' households, and nearly all of them were so listed at the later date[15].

The criterion of 'current costs' thus performs well as an explanation of the observed pattern of fertility reduction and we shall consider the implications of this in the concluding section. At this point, however, we must try to answer the second of the two questions posed earlier – that of finding some explanation for the timing of the reduction in fertility. The available information on the socio-economic position of the study population is meagre, but both the occupational and residential data contained in the vital registers suggest that the major changes in these respects did not occur until around the turn of the century (see note 3).

Quakerism as a whole underwent a series of reforms in the 1760s and it might be argued that this so-called 'restoration of the discipline' implied a collective re-definition of Friends as a community under stress. This process should certainly have strengthened any features of the discipline, such as the system of poor-relief, which tended toward fertility reduction, but it is not itself a sufficient explanation since the reconstitutions conducted by Eversley (1981) display no trace of a decline in the fertility of Quaker communities elsewhere in the British Isles (see Figure 8.1).

A more straightforward explanation, which might also account for the regional variation among British Quakers, is the reduction in mortality from the very high levels detected among the study's earlier cohorts (see Figure 8.8)[16]. Lower fertility might come about as a direct 'homeostatic' response by parents to the enhanced survival prospects of their children, but it has recently become fashionable to see the process as mediated by a complex of far reaching changes in attitudes and relationships within the family. The argument for a 'discovery of childhood' at some point in the eighteenth century is hard to evaluate, not least because its proponents are

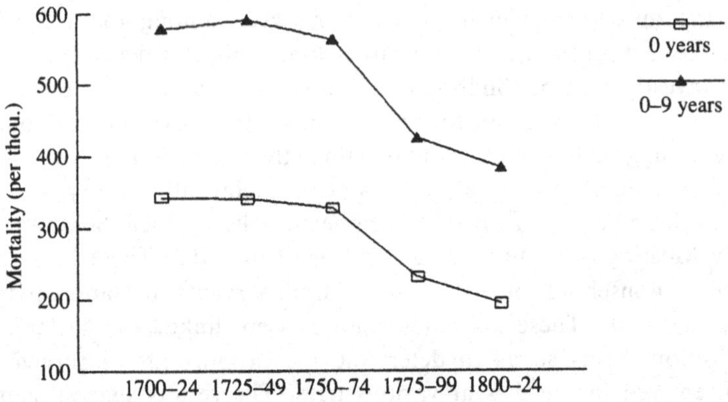

Figure 8.8. Life table mortality probabilities, London Quakers.

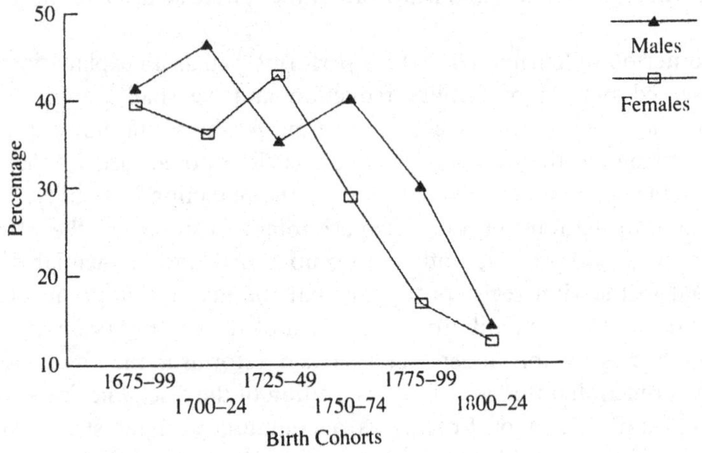

Figure 8.9. Incidence of 'replacement' naming as a percentage of total 'opportunities'.

apt to take the general lack of evidence before then as mute testimony on their behalf. This difficulty applies to Friends as much as to the wider society (Frost, 1973: p. 1973).

There is, however, one continuous source to which appeal has sometimes been made. This is the structure and content of personal names. Writers such as Lawrence Stone have claimed that the practice of naming infants after deceased siblings reflected a low level of individuation in their parents' eyes and that the demise of this practice, together with the use of multiple names, was due to a corresponding awareness of the unique personality of the child (Stone, 1979: pp. 257–8). In the present study a marked change

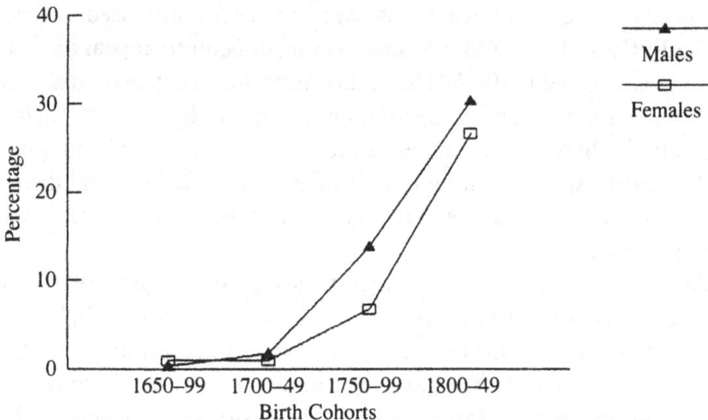

Figure 8.10. Incidence of middle names as a percentage of entries in birth register.

was detected, in both these respects, after the middle of the eighteenth century.

The entries in the birth register were analysed so as to measure the frequency of middle names, whilst the family reconstitution forms were examined to determine the proportion of all cases in which the name of a deceased child was 'available', in which parents actually bestowed this name on a new born sibling. The results (see Figures 8.9 and 8.10) are consistent with the hypothesis. The frequency of 'repeat' naming (Figure 8.9) declines earlier among girls than among boys, but boys were more likely to have a middle name than were girls (Figure 8.10). The major reduction in 'repeat' naming, taking both sexes together, coincides with the substantial reduction in mortality in the 1770s.

Conclusion

A reduction in mortality, associated with a change in the way parents responded to their children, may account for the fall in fertility after 1750. The link between onomastic patterns and social mentality is, however, a mere hypothesis and more evidence of a qualitative character would be required before a claim of this nature could be firmly substantiated. Similar considerations apply to the question of 'modern' and 'traditional' mentalities. We have not found it necessary to invoke modernization as a component of fertility change and this notion becomes increasingly doubtful if we accept the ambiguous evidence for moderate fertility control earlier in the eighteenth century.

The claim that England was ever a 'traditional' society, in the sense in which this term is generally used, has been extensively criticised in recent years (MacFarlane, 1978) and it is, in any case, difficult to appeal to such a transition as late as the 1750s. We have, however, made extensive use of the concept of 'rational' fertility behaviour advanced by writers such as Knodel. Our aim here was to challenge the assertion that such behaviour is confined to parity-specific family limitation and that the technical distinction between this and 'natural fertility' corresponds to a wider socio-cultural dichotomy.

The flaw in this argument, we suggest, lies in its equation of fertility regimes exclusively with the observable patterns of reproductive behaviour to which they give rise, at the expense of the social and economic processes underlying such behaviour. We have shown, through the construction of a very simple model, that the same criterion of decision making will result in either parity-specific, or parity-independent behaviour depending on the values of certain other variables. As such we have been concerned to explore the 'logic of the possible', rather than to argue that the limited case of the London Quakers is representative of some larger universe. Our findings do nonetheless have some wider implications.

The criterion on which the model is based is a predominant concern to minimize the current cost of children over a finite period of dependency, rather than a preoccupation with the problems of inheritance or 'heirship'. Such a concern may be appropriate for a range of historical populations, particularly in urban contexts, or for those societies outside this area in which ownership of land is not in individual hands. The 'current cost' criterion gives rise to a 'spacing' strategy where the period of childhood dependence is short, or where marriage is delayed into the late twenties. The second of these conditions was found historically in parts of northern Scandinavia, whilst the first may be relevant to a number of contemporary Afro-Asian populations which exhibit lengthy birth intervals (Caldwell & Caldwell, 1981).

This is not to imply that such intervals are 'really' a form of conscious fertility strategy, in all cases, rather than a means of promoting child health or a ritual observance. Our argument is rather that 'the same' demographic behaviour may have very different meanings in different social or economic contexts. Thus spacing behaviour which is begun as a means of reducing infant mortality may take on other functions in changed circumstances, without an external discontinuity being visible. Something of this kind may have occurred among some of our reconstituted families in the course of the eighteenth century.

In the same way, we should beware of treating reductions in marital fertility and delayed marriage as simple alternatives to each other, for variations in age at marriage may introduce qualitative changes in the options for fertility control. The specific implications of our findings for the population of London in general are less clear but a number of points deserve further investigation. The capital apparently experienced a sharp reduction in infant mortality in the later eighteenth century and the additional surviving children must have imposed considerable strains on many households, not least on the available living space.

Seventeenth century London had exhibited the characteristic signs of a 'high pressure' demographic regime, but many parts of the capital had low marital fertility by the 1860s. The material circumstances of the population would, we suggest, have been such as to make 'current costs' a more appropriate criterion than final parity for many families. The part played by lengthening birth intervals in fertility change over this period may thus be worthy of as much attention as has hitherto been devoted to the passage from natural fertility to family limitation[17].

Acknowledgements

The family reconstitution study whose results are reported in this paper was carried out while the author was a research student in the ESRC Cambridge Group and I should like to thank the staff of the Group, especially Professor E. A. Wrigley and Dr. R. S. Schofield, for their help and advice at all stages of the project. I should also like to thank the staff of the Friends' House Library, Euston Road, particularly Edward Milligan and Malcolm Thomas, whose generosity with their time and encouragement was essential to its success.

Notes

1 France, as is well known, is an apparent exception to this generalization, fertility declines being evident in many areas from the late eighteenth century. Here too, however, a second phase of fertility reduction is evident from the last quarter of the nineteenth century (see Wrigley, 1985).

2 See Coale & Watkins (1986) for a summary of results. For an important caution against the tendency to treat European fertility decline as a unitary phenomenon with a single explanation see Woods (1987).

3 Of the two meetings one, Southwark, took in the built up area south of the Thames, whilst the other, 'Peel', covered the inner suburban parishes lying outside the city wall running roughly from Fleet Street clockwise to Aldersgate. The occupational data in the registers suggest that, for most of the eighteenth century, the study population was predominantly one of artisans and small tradesmen. After 1800 there is evidence of an increasing proportion of professional, entrepreneurial and banking occupations, and of a geographical shift towards the new outer suburbs. Quaker vital registers were modelled on their Anglican counterparts, although they record birth rather than baptism and are generally more informative. The nature of Quaker organization and record keeping is discussed in Landers (1984) Chapter 3, whilst for a more detailed description of the study population see Landers (1990 in press).

4 For an outline of this method as applied to English data see Wrigley (1966a). Henry's conclusions were first reported to an English speaking audience in Henry (1961).

5 The concept itself was first developed in the early 1950s (Henry, 1953), but first presented in a general context in the key article published in 1961 (Henry, 1961). Recently, natural fertility has been the subject of an entire IUSSP symposium volume (Leridon & Menken, 1979). See also Wilson *et al.* (1988).

6 The ensuing discussion of this question is based primarily on the writings of John Knodel and Etienne Van de Walle rather than on those of Henry himself, since the latter makes few explicit remarks on the broader socio-cultural correlates of natural fertility or family limitation. We would argue, however, that a view close to that of Knodel's is implicit in Henry's 1961 article, the roots of variation in natural fertility being assumed to lie in physiological factors (pp. 90–1) or in social mentality rather than in economic or material circumstances.

Thus in speaking of 'social' factors affecting natural fertility, he alludes only to 'sexual taboos, for example during lactation' which 'cannot be considered a form of birth control' because of their parity-independent effect (p. 81). The implicit importance attached to social mentality in the transition to family limitation is indicated by the observation that populations likely to be practising natural fertility are identified on the basis of 'a hypothesis based on time, place, thought, culture, type of society etc.' (*ibid*).

In present day non-European societies Henry thought that the evidence for the prevalence of natural fertility was reinforced 'by studies showing that the very idea of limiting births is foreign to them' (*ibid*). Perhaps the most interesting, because unintentional, indication of the determining role which Henry attributed to 'culture' in the fertility of pre-transitional societies is his remark that fertility data, obtained from an Iranian village survey 'seem to be acceptable since, when applied to Muslim populations of the Mediterranean coast, they give birth rates which have been observed or evaluated by other means' (p. 88).

7 Knodel acknowledges the possibility that birth intervals may be extended 'for the benefit of both the child's and the mother's health' and that 'what has sometimes been interpreted by outsiders as a "taboo" ... is viewed by the indigenous populations as a deliberate and rational attempt to space their births' (1977: p. 220; see also Knodel, 1988: pp. 318–34). Here as elsewhere, however, he stresses (cf. Knodel & Van de Walle (1979)) the implicit distinction between 'rationality' in the choice of appropriate means to a given end and 'rationality' in the formulation of the goals themselves. The author is, in addition, equivocal as to the nature and prevalence of the 'health' motive:

> Even in situations where the fertility reducing effect of breast-feeding or postpartum abstinence is both recognised and desired, they ... are typically deeply entrenched in the *mores* of the populations. Because of strong normative pressures to follow customary practices, more of an act of volition may be required not to breast-feed a child or not to abstain during the postpartum period rather than to act in accordance with the prescribed custom.

(Knodel, 1977: p. 241.)

8 See Cleland & Wilson (1987) for a recent defence of the distinction, and the accompanying 'ideational' interpretation of the fertility transition, together with a review of some contrary positions.

9 See for instance the African total fertility rates assembled in Lesthaeghe (1986), as compared with the European reconstitution results collected by Flinn (1981, Appendix table 1). For a discussion of such variations in the context of varying systems of family, kinship and community support see Smith (1981) and Lesthaeghe (1980).

10 For a review of these arguments as they relate to demographic theories of fertility decline see Woods (1982: pp. 159–84).

11 A number of marital fertility measures can be evaluated on the basis of the family reconstitution material. Each has its own particular uses, but the rules of observation used in the calculations are not always identical and so the universe on which each series is defined may differ slightly.

12 The fecundability calculations were also carried out using distributions of protogenesic intervals 'trimmed' at 36 months. The results are given in the table:

Fecundability estimates

Age at marriage	*1650–1749*	*1750–1849*
–24	0.26	0.35
25–34	0.35	0.16

The results indicate a generally higher level of fecundability, as we would expect, but the overall pattern remains unchanged.

13 This is particularly true in the case of France. See Wrigley (1985).

14 The construction of confidence intervals for marital fertility parameters is a complex question and a large number of observations are required if useful results are to be obtained (Wilson, 1982: Appendix 2; Wilson *et al.*, 1988). In the present study a simpler technique was employed. The data were pooled into two cohorts, according to whether the date of marriage was before or after 1750, and analysed by means of 2×2 contingency tables. The results were generally encouraging. At the lower parities the inter-cohort difference in the frequencies of 'long' intervals among women marrying above 24 was significant at the 0.1% level, whilst at the higher parities the difference was significant at the 5% level for the younger brides. In the second cohort the difference in the frequency of long birth intervals, at low-parities, between the two bridal age-groups was also significant at the 5% level. The difference in fecundabilities between the two age-groups after 1750 was tested indirectly by examining the proportion of protogenesic intervals falling in the 9–11 month range. Here again the difference was significant at the 5% level. In both cohorts the difference between the proportions observed among the 25+ year age-group was significant at the 5% level.

15 The interpretation of household listings of this kind has given rise to controversy; for an authoritative discussion of the conceptual and methodological problems arising see Wall (1983). In the present case the purpose for which the documents were drawn up was such that all persons listed would have had a claim on the resources of the Meeting. It is also highly likely that, with a few limited exceptions, all those so listed were actually resident within the district. Servants and apprentices gained a settlement in their masters' meeting and the Southwark lists include members of both these categories. Young people who left home for service elsewhere would thus be 'double counted' if they were to be included as 'chargeable' both to Southwark and to the meeting in which they resided. Such a position seems highly unlikely given the extreme care with which meetings avoided excess liabilities in respect of relief.

16 For a detailed analysis of the mortality data from the reconstitution study see Landers (1990 in press). Mortality patterns in London as a whole during the period considered here are discussed in Landers (1987), and Landers & Mouzas (1988).

17 See Crafts (1989) for recent evidence concerning the importance of birth spacing in the English national fertility decline.

References

Byers, E. (1982). Fertility transition in a New England commercial center: Nantucket, Massachusetts. *Journal of Interdisciplinary History*, **13**, 17–40.

Cain, M. (1982). Perspectives on family and fertility in developing countries. *Population Studies*, **36**, 159–75.

Caldwell, J. C. (1976). Towards a restatement of demographic transition. *Population and Development Review*, **2**, 321–66.

Caldwell, J. C. (1982). *The Theory of Fertility Decline*. London: Academic Press.

Caldwell, P. & Caldwell, J. C. (1981). The function of child-spacing in traditional societies and the direction of change. In *Child Spacing in Tropical Africa*, ed., H. Page & R. Lesthaeghe. London.

Carlsson, G. (1966). The decline of fertility: innovation or adjustment process? *Population Studies*, **20**, 149–74.

Cleland, J. & Wilson, C. (1987). Demand theories of fertility and the fertility transition: an iconoclastic view. *Population Studies*, **41**, 1, 5–30.

Coale, A. J. (1973). The demographic transition reconsidered. In *International Population Conference, Liège 1*, pp. 53–72. Liège: IUSSP.

Coale, A. J. & Trussel, T. J. (1974). Model fertility schedules: variations in the age structure of childbearing in human populations. *Population Index*, **40**, 185–258.

Coale, A. J. & Trussel, T. J. (1978). Technical note: finding the two parameters that specify a model schedule of marital fertility. *Population Index*, **44**, 203–13.

Coale, A. J. & Watkins, S. (eds.) (1986). *The Decline of Fertility in Europe*. Princeton: Princeton University Press.

Crafts, N. F. R. (1989). Duration of marriage, fertility and women's employment opportunities in England and Wales in 1911. *Population Studies*, **43**, 325–35.

Demeny, P. (1972). Early fertility declines in Austria-Hungary: a lesson in demographic transition. In *Population and Social Change*. D. V. Glass & R. Revelle. London: Edward Arnold.

Dupâquier, J. & Lachiver, M. (1969). Sur les débuts de la contraception en France ou les deux malthusianismes. *Annales: Economies, Societies, Civilisations*, **24**, 1391–406.

Elliott, V. B. (1981). Single women in the London marriage market: age, status and mobility, 1589–1619. In *Marriage and Society*, ed. B. Outhwaite. London: Europa.

Eversley, D. E. C. (1981). The demography of the Irish Quakers 1650–1850. In *Irish Population, Economy and Society*, ed. J. M. Goldstrom & L. A. Clarkson. Oxford: Clarendon Press.

Finlay, R. A. P. (1981). *Population and Metropolis*. Cambridge: Cambridge University Press.

Flinn, M. W. (1981). *The European Demographic System 1500–1820*. Brighton: Harvester Press.

Frost, J. W. (1973). *The Quaker Family in Colonial America*. London: St. Martin's Press.

Henry, L. (1953). Fondements théoriques des mesures de la fécondité naturelle. *Revue Institut International de Statistique*, **21**, 131–51.

Henry, L. (1961). Some data on natural fertility. *Eugenics Quarterly*, **8**, 81–91.

Henry, L. (1976). *Population – Analysis and Models*. London: Edward Arnold.

Knodel, J. (1977). Age patterns of fertility and the fertility transition: evidence from Europe and Asia. *Population Studies*, **31**, 219–49.

Knodel, J. (1988). *Demographic Behavior in the Past*. Cambridge: Cambridge University Press.

Knodel, J. & Van de Walle, E. (1979). Lessons from the past. Policy implications of historical fertility studies. *Population and Development Review*, **5**, 217–45.

Knodel, J. & Wilson, C. (1981). The secular increase in fecundity in German village populations: an analysis of reproductive histories of couples married 1750–1899. *Population Studies*, **35**, 53–84.

Landers, J. (1984). Some Problems in the Historical Demography of London 1675–1825. Unpublished University of Cambridge Ph. D. thesis.

Landers, J. (1987). Mortality and metropolis: the case of London 1675–1825. *Population Studies*, **41**, 59–76.

Landers, J. (1990). Mortality levels in eighteenth century London: a family reconstitution study. In *Living and Dying in London*. ed. R. Porter. London: Croome Helm (in press).

Landers, J. & Mouzas, A. J. (1988). Burial seasonality and causes of death in London 1670–1819. *Population Studies*, **42**, 59–83.

Leridon, H. & Menken, J. (1979). *Natural Fertility*. Liege: Ordina Editions.

Lesthaeghe, R. (1980). On the social control of reproduction. *Population and Development Review*, **6**, 527–48.

Lesthaeghe (1986). On the adaptation of sub-Saharan systems of reproduction. In *The State of Population Theory*, ed. D. Coleman & R. S. Schofield. Oxford: Blackwell.

Livi-Bacci, M. (1977). *Two Centuries of Italian Fertility*. Princeton.

Livi-Bacci, M. (1986). Social-group forerunners of fertility control in Europe. In *The decline of fertility in Europe*, ed. A. J. Coale & S. Watkins.

MacFarlane, A. (1978). *The Origins of English Individualism: The Family, Property, and Social Transition*. Oxford: Blackwell.

Schofield, R. S. (1985). English marriage patterns revisited. *Journal of Family History*, **10**, 2–20.

Schofield, R. S. (1989). Family structure, demographic behaviour and economic growth. In *Famine, Disease and the Social Order in Early Modern Society*. ed. J. Walter & R. S. Schofield. Cambridge: Cambridge University Press.

Smith, R. M. (1981). Fertility, economy and household formation in England over three centuries. *Population and Development Review*, **7**, 595–622.

Stone, L. A. (1979). *The Family, Sex and Marriage in England 1500–1800*. London: Weidenfeld & Nicolson.

Wall, R. (1983). Introduction. In *Family Forms in Historic Europe*, ed. R. Wall, J. Robin & P. Laslett. Cambridge: Cambridge University Press.

Wells, R. V. (1971). Family size and fertility control in eighteenth century America: Quaker families. *Population Studies*, **25**, 73–82.

Wilson, C. (1982). *Marital Fertility in Pre-industrial England 1550–1849*. Unpublished University of Cambridge Ph. D. thesis.

Wilson, C., Oeppen, J. & Pardoe, M. (1988). What is natural fertility? The modelling of a concept. *Population Index*, **54**, 4–20.

Woods, R. I. (1982). *Theoretical Population Geography*. London: Longman.

Woods, R. I. (1987). Approaches to the fertility transition in Victorian England. *Population Studies*, **41**, 283–311.

Woods, R. & Smith, C. W. (1983). The decline of marital fertility in the late nineteenth century: the case of England and Wales. *Population Studies*, **37**, 207–25.

Wrigley, E. A. (1966a). Family reconstitution. In *Introduction to English Historical Demography*. ed. E. A. Wrigley. London: Weidenfeld & Nicolson.

Wrigley E. A. (1966b). Family limitation in pre-industrial England. *Economic History Review*, **19**, 82–109.

Wrigley, E. A. (1978a). Marital fertility in seventeenth century Colyton: a note. *Economic History Review*, **31**, 429–36.

Wrigley, E. A. (1978b). Fertility strategy for the individual and the group. In *Historical Studies of Changing Fertility*. ed. C. Tilly. Princeton: Princeton University Press.

Wrigley, E. A. (1985). The fall of marital fertility in nineteenth century France (part one). *European Journal of Population*, **1**, 31–60.

Wrigley, E. A. & Schofield, R. S. (1981). *The Population History of England 1541–1871: A Reconstruction*. London: Edward Arnold.

9 *Population, growth, innovation and resource exploitation*

E. BOSERUP

The most important link between population growth and economic development is provided by the reductions in communication and transport costs which follow from increases in population density. Where population size is small improvements in labour productivity are impeded because limited numbers restrict the possibilities for labour specialization outside the family. Specialized producers of consumer goods, agricultural equipment, and services cannot earn a living where villages are small and widely scattered, but as population density increases, and villages become larger and more closely spaced, so the market becomes big enough to support a variety of different types of artisans and performers of specialized services. This development is crucially important because it is only specialized producers who can draw the full benefits of learning by doing, or make economic use of specialized equipment and specialized, full-time, training and education. Thus, agricultural productivity increases as some of the inputs improve in quality, and the advantages so gained are shared by agricultural and non-agricultural producers (Boserup, 1988).

Increases in population allow transport facilities to be improved, and costs to be reduced, because of the growth in the local labour force and in the volume of demand. When regional population densities pass a certain threshold, villages may become linked together by road networks giving them access to a wider market at lower costs. Unless the area is situated a very long way from more developed regions, communications with the outside world will also become more frequent, facilitating the transfer of technology. But the creation of transport facilities, like other forms of investment in rural infrastructure, depends on the decisions of local leaders or central government. Analyses of the economic effects of population growth which fail to take account of government policies are thus quite inadequate.

Improved transport facilities also promote urbanization, because towns can now obtain supplies more cheaply and from a larger area, and because they gain access to a wider market. When specialized producers of non-agricultural products and services congregate in urban centres, further productivity increases are obtained by additional specialization and communication between them, and by economies of scale in physical and human infrastructure investment. Ease of communication between specialists working in the same field, and between people with different specializations, together with the presence in towns of educational establishments and training centres, creates an atmosphere favourable to innovation, whether by technology transfer or invention.

Inventions in both science and technology require communication and cooperation between people doing the same sort of work and facing the same sort of problems, such as the growing scarcity of resources brought on by increases in population. Moreover, the possibility of importing materials to substitute for exhausted local resources, and improving the use of local resources by technology transfer, are both enhanced as contacts are developed with more remote communities. The declining availability of physical resources per capita may thus be much more than outbalanced by the increasing productivity of human resources resulting from the mutually reinforcing processes of improvements in communication and infrastructure, growing specialization of labour, market expansion, improvements in technique, resource substitution and further innovation (Boserup, 1981).

Effects of technological innovation on fertility and savings

Population growth, accompanied by innovation and resource substitution, leads to change in the structure of the economy. More and more items of consumption are produced by wage labour in specialized enterprises, which replace or supplement those using family labour. As the production of goods and services by families, for their own use, is replaced by the consumption of items produced elsewhere, the economic role of the family declines and so does fertility and the rate of population growth (Boserup, 1986).

One of the roles of the family is to produce offspring. We should treat as investment in human capital not simply the cost of educating children in schools and universities, but all the costs inherent in child rearing. This was in fact the general attitude throughout most of human history. Mortality rates were so high that both parents and community leaders had a strong interest in high fertility. Parents needed surviving children to assist them in their work and to support them in case of temporary disability, whilst the community needed a large younger generation for defence, and to undertake construction works or pay the taxes necessary to cover public expenditure. In order to secure high fertility, women were discouraged, or prevented, from participating in any activity that might compete with their reproductive role, and they were socialized to equate female status with high fertility. Most religions reflect this attitude (Lutz, 1987).

The modern economic theory of the family explains fertility behaviour at different stages of economic development in terms of the rising net costs of rearing children, a major role being here played by the obstacles which child care places in the way of married women taking paid employment (Becker, 1981). But the theory fails to recognize that maximization of income is not always the parents' first priority. Where this comes into conflict with the goal of survival, whether in the short run or in old age, it is the latter which will be preferred. If low fertility is inimical to survival, then people will marry early and have large families even if the effect on their income is likely to be negative. In many developing countries, with widespread contraception, fertility surveys reveal that desired family size is around four children (Lightborn & MacDonald, 1982). This seems to imply that, despite the rising costs of child rearing and the declining use of child labour, parents will still want to be sure of having at least one surviving son to depend on when they are no longer able to support themselves.

As subsistence production declines, so more and more of the family members must become wage earners, or specialized producers of goods and services, so that the family can purchase items of consumption, or pay the taxes nexessary for the public delivery of specialized services. Thus in the course of time some family members switch from subsistence activities to money earning, whilst others specialize in domestic activities and child care which are unpaid. When the family comes to rely on purchase, or public delivery, for such things as water and energy supply, education, health services, and care of the old, then its adult male members, together with children and unmarried females, must become either money earners or participants in institutionalized education and training. At the end of this process the mother is left to take care of the small children in a less and less labour intensive household.

From being a part time job, shared by many family members, child care has become the major activity of the mother, and the conflict between family reproduction and women's work and career becomes acute. Moreover, the transfer of an increasing share of social services from the family to public, or private, institutions reduces the importance of offspring for the survival of the parents, and thus their motivation for having children. In countries where even the poor have old age pensions, health services, and a minimum support from public funds when they would not otherwise be able to survive, fertility levels are close to, or below, replacement.

Population growth accompanied by technological innovation influences not only fertility rates, but also activity rates and the ratio of savings to consumption (Kelley, 1988a, b). In families with high dependency ratios children start work earlier, and the adults work longer hours, than happens where there are few or no children to support (King, 1987; Ministry of Social Affairs, 1980). Moreover, the rate of savings from a given income depends not only on the size of that income, as is usually assumed, but on the strength of the motivation to save. People with low incomes, who value a large family above the enjoyment of leisure and luxury, may have high savings rates, and those who depend on a family enterprise for support must save if the enterprise is to expand as family size increases, and the children arrive at working age. Wage earners do not need to save in order to carry on a business, but in poor countries, where they depend on their family for survival in old age and other emergencies, they are likely to restrict their consumption standards so as to invest in rearing children who grow up to become wage earners themselves and support their parents.

In order to survive the competition owner-operated family enterprises with hired labour may have very high savings rates. But the motivation for savings is lower for the salaried managers, who supplant the owner-operators as corporately owned concerns begin to out-compete owner-operated enterprises. The economic survival of the managers does not depend on the size of their personal savings, but on their ability to expand the business by efficient use of credits. And a high level of luxury consumption is a means of showing that they are successful managers, qualified for a managing job in a larger concern.

When social security systems improve, and are extended to all groups in the population, not only fertility but also the motivation for personal savings declines, especially in countries which provide both elementary and more advanced education free of cost to the parents. Instead of saving, wage and salary earners in the modern welfare state exert political pressure on governments in favour of increased social security expenditure, and reduced taxes, with the result that public deficits make inroads into the savings potential. The reductions in private expenditure, which parents achieve by having smaller families, are likely to be offset by higher personal consumption rather than higher savings.

Population growth and resource exploitation

Population growth reduces the per capita stock of non-human, natural resources, if these are defined so as to include free and non-used resources. Thus if population growth is sustained over a long period it becomes necessary to utilize such resources more intensively, and some hitherto 'free' resources become scarce, so that they must either be protected or replaced by others. Intensification of land use occurs automatically as population grows. When numbers become too large for the older extensive systems of land use fallow is first shortened and then eliminated, and pastures and forests are replaced by fields.

This process of intensification is accompanied by changes in the composition of the human diet. The meat of animals which require a great deal of land gives way to that of animals needing less land, and animal foods are replaced by vegetables. But if crop yields are to be maintained, and damage to the land avoided, additional labour inputs are necessary both for current operations and for investments in land improvement and ancillary infrastructure (Boserup, 1965). All societies which have had to feed a multiplying population have employed this sequence of gradual transformations in agriculture, with increasing inputs of both labour and capital, whether in cash or in kind.

When the survival of individuals and communities was dependent on intensifying agriculture without benefit of industrial inputs, as was the case until recently in much of Asia, diminishing returns per capita could only be avoided by increasing the labour inputs of all family members: yielding a subsistence income earned by the very hard toil of children and adults alike. Populations which are able to import knowledge and improved equipment from elsewhere are better placed to avoid diminishing returns to labour, but technical changes obtained by imports of knowledge and other inputs are not a free gift, since they require the investment of human and physical capital.

It is often forgotten that both technology transfer and invention require previous investments in economic and social infrastructure. Lack of a sufficient stock of human and physical capital has proved a serious handicap to the transfer of technology to many developing countries. Since nearly all types of infrastructure display economies of scale, and require a minimum population size and density to be economical, many thinly peopled regions have failed to benefit from this source (Boserup, 1981).

Human history provides many examples of peoples who degraded their natural resources by over-exploitation and left their habitats a desert, but we also have abundant evidence of those who matched population growth with investments in the improvement and protection of land and water resources, and by substituting new sources of energy and materials for those which had become scarce.

New technology may replace a non-renewable resource which has become exhausted with one which is still abundant, but technological changes of this kind often involve the replacement of over-exploited renewable resources by ones which are non-renewable (Kneese, 1988). Energy consumption, once derived from wood, flowing water, and animal power, now depends more and more on fossil fuels and uranium because of the rise in world population. In monetarized economies such substitution is promoted by relative increases in the price of materials threatened with exhaustion, which provide the motivation for technology transfer or invention. However, the introduction of new technology by either of these routes can also have negative effects for the entire population, and not just for those groups which have retained the traditional methods. Some inventions damage the physical or biological environment, whilst others threaten human health and survival, and some do both. Examples of such negative effects are numerous, and have sometimes led to the decimation of populations and the collapse of civilizations.

Many irrigation techniques have damaged the land through increasing salinity, or destroyed human life and health by flooding, or the spread of water borne diseases. Agricultural chemicals have preserved soil fertility but polluted water resources. Improved transport technology helps to avoid famine, but also spreads disease, and increased urbanization and crowding in urban centres has promoted epidemics and digestive and airborne disease. Substitution of timber by charcoal-produced iron depleted the forests, whilst the later substitution of coal–iron technology for this method, and of fossil fuel for fuel wood, created air pollution and promoted tuberculosis and other respiratory diseases. The substitution of atomic energy, in turn, reduced these threats to human life but brought other hazards in its train.

The negative consequences of technological change have sometimes forced the abandonment of settlements, both in historical times and more recently, but in other cases they led to an acceleration of the process by stimulating inventiveness (or technology transfer), so as to ameliorate the damage caused by earlier innovations (Boserup, 1981). Where human health is concerned, the net effect of technological change has been strongly positive until now, as witness the decline of mortality and morbidity, but in the case of the environment it has often been negative. Threats to human life and health are always taken more seriously than threats to the physical environment, and it is only where these seem to be life threatening that the financial means to combat them effectively are forthcoming. In recent decades the development of dangerous chemicals and atomic power has focussed public opinion on the negative effects of innovation.

It was mentioned above that a continuing process of innovation ends by reducing fertility, and may eventually arrest population growth. But even if such growth is arrested further innovation is still required to avoid the negative effects of intensive resources utilization, and to replace non-renewable resources by others as they become exhausted. Further improvements in the quality of human resources will be necessary to assure continuing technological change. The promotion of new innovations, and the development of measures to offset the negative effects of innovations past and future, will continue to be urgent problems.

It is often possible to avoid, or substantially reduce, the risks that a new technology will have negative effects by incurring additional costs, either through more extensive testing or by ancillary investment and current expenditure. Such measures are often avoided, because the producer either benefits from being first in the field with an invention, or thereby becomes competitive in national and international markets. If the negative effects are not discovered until after – perhaps long after – the technology is put into practice, the producers are likely to have a vested interest in continuing with methods of production which may be polluting or dangerous, whilst consumers may object to paying higher prices for cherished commodities.

The costs involved in mitigating such negative effects are often public costs, financed fully, or partly, by general taxation. When this is the case, the dangerous or polluting products and services acquire a competitive advantage against those that have no need of special measures to stop them damaging health or the physical environment. In such circumstances, where government expenditure is required, the damage can be limited or avoided by a levy which passes the full cost of the polluting, or dangerous, activities onto the producers themselves. In this way the price mechanism can be used to encourage the substitution of better methods of production and less dangerous patterns of consumption.

However, if a government prohibits undesirable technologies, or imposes levies on the producers so as to limit their market or finance appropriate environmental measures, that country may lose international competitiveness to others which spend less on protective measures and controls. This is a very important deterrent to such action at the national level, because of the large and increasing importance which international markets have acquired for many producers.

Most technological inventions are made in the research departments of the large multinational companies. Only very large enterprises can afford the substantial investments in research, specialized training, and other infrastructure, which are needed if further innovation is to be achieved through inventions. If a small company happens to make a promising invention, it is usually taken over by one of the large multinationals. National governments have little scope for controlling new inventions made by the multinationals, who can circumvent measures against dangerous or polluting processes by placing production in those countries which have the fewest regulations, thus augmenting the resistance against such controls elsewhere.

However, political pressure for internationally coordinated action is gaining strength as evidence accumulates of the kind of resource destruction which cannot be contained within national frontiers. Popular protest is becoming intense because a threat is now perceived not merely to income, but to health and survival, and these are seen to depend on the action or inaction of people in neighbouring countries, or even those on other continents.

The issue is most burning in Europe, where a large number of high technology countries crowd together in a small subcontinent. It has created support for the efforts of the European Community to coordinate government policies and reduce discrimination in the European market. It is increasingly acknowledged that European environmental problems are now so serious that closer limits to the sovereignty of individual nations have become necessary. But many environmental and health problems have a global basis, and international coordination is bound to proceed further. Strict adherence to the principle of national sovereignty is no longer compatible with the use, and further development, of high level technologies.

References

Becker, G. (1981). *A Treatise on the Family*. Cambridge Mass.: Harvard University Press.

Boserup, E. (1965). *The Conditions of Agricultural Growth*. London: Allen and Unwin.

Boserup, E. (1981). *Population and Technology*. Oxford: Blackwell.

Boserup, E. (1986). Shifts in the determinants of fertility in the developing world. In *The State of Population Theory*, ed. D. Coleman & R. S. Schofield. Oxford: Blackwell.

Boserup, E. (1988). Population growth as a stimulant to agricultural development. In *Probleme und Chancen Demographischer Entwicklung in der Dritten Welt*, ed. G. Steinmann *et al.* Berlin–London: Springer Verlag.

Kelley, A. (1988a). Economic consequences of population change in the Third World. *Journal of Economic Literature*, **XXVI**, December, 1685–728.

Kelley, A. (1988b). Population pressures, savings and investment in the Third World: some puzzles. *Economic development and culture change.*

King, E. (1987). The effect of family size on family welfare: what do we know? In *Population Growth and Economic Development: Issues and Evidence*, ed. D. Gale Johnson & R. Lee. Madison: University of Wisconsin Press.

Kneese, A. (1988). The economics of natural resources. *Population and Development Review*, supplement to vol. **14**.

Lightborn, R. & MacDonald, A. (1982). *Family Size Preferences*. World Fertility Survey, Comparitive Studies no. 14.

Lutz, W. (1987). Culture, religion and fertility: a global view. *Genus*, **XLIII**, 3–4.

Ministry of Social Affairs (1980). *Economics and Work Conditions of Families with Children*. Report no. 1 from the Commission on Children, Copenhagen (in Danish).

10 *Fertility decline in developing countries: the roles of economic modernization, culture and Government interventions*

JOHN CLELAND

Introduction

As described by Allan Hill in this volume, the last twenty years have witnessed spectacular declines in human fertility. In 1970, birth rates were high and showed little sign of change in any major developing region of the world. The few exceptions were clearly idiosyncratic, comprising small island states such as Fiji and Mauritius, the city-states of Hong Kong and Singapore and larger territories that had received massive North American aid, as in the case of the Republic of Korea and Taiwan.

Some twenty years later the position is very different. Fertility has declined throughout Central and South America and appears to be approaching replacement level in such countries as Brazil. In Asia, there is greater variation in recent trends. Steep declines have occurred in China, Indonesia and Thailand. More modest falls have been registered in Malaysia, the Philippines, Vietnam and India. In Bangladesh there is recent evidence of a moderate decline, while in Pakistan fertility may even have risen.

In the predominantly Arabic belt of North Africa and the Middle-East, birth rates remain high but downward trends are well established in Egypt and Tunisia and there have been more recent falls in Morocco and Algeria. Similarly, in sub-Saharan Africa, high fertility persists and several countries have experienced a rise rather than a fall over recent decades. Zimbabwe and Botswana are the only countries in this region with unmistakeable trends to lower fertility, but Kenya is about to join their ranks.

Attempts to account for these trends are dominated by economic considerations. Esther Boserup has presented in this volume many of the central arguments. In traditional societies, large numbers of children are needed for family labour. They provide further utility as a source of security in old age and against other hazards. Child rearing is not incompatible with other activities and therefore the opportunity costs of children are low. Fear of childhood mortality may further buttress pro-natalist sentiments.

The economic modernization of societies progressively strips away these utilities of children. Family enterprises are replaced by wage employment and alternative forms of security arise. Both the direct and indirect costs of children increase with the advent of educational expenses and rising employment opportunities for women outside the home. Under this onslaught of changes that undermine the economic rationality of large families, fertility inevitably falls.

These arguments are so plausible and so deeply entrenched in all the dominant theories of demographic transition that they are rarely challenged. The purpose of this chapter is to mount such a challenge by demonstrating that economic determinism cannot provide a convincing explanation for recent Third World fertility trends and that cultural factors, encompassing knowledge, beliefs and values, assume a far greater importance than is generally realized. In the course of the discussion, the influence on fertility of government interventions, in the form of family planning programmes, will also be considered.

The inadequacy of economic explanations

The case for a more cultural and a less exclusively economic framework for understanding fertility transition can be argued negatively by pointing out the inadequacies of economic explanations, and positively by demonstrating the links between cultural units, such as ethnicity and religion, and reproductive behaviour. This section presents the negative or residual evidence, by means of a critical examination of four themes that are central to economic theorizing about human fertility. These themes are macro-economic development, modes of production, female employment and the role of children as forms of family security.

The search for a clear-cut relationship between macro-economic development and fertility change is longstanding. It remains true, of course, that all highly industrialized countries have low fertility and that most of the poorest countries in the world are still characterized by high fertility. However, attempts to relate the onset and speed of fertility decline to stages of economic development have been unsuccessful. At the time of the European fertility transition between 1880 and 1930, there was very wide variation in economic circumstances. England for instance, was highly industrialized and urbanized while Bulgaria was a poor country with an economy still dominated by subsistence agriculture. Yet within a relatively short span of time, all European countries (with the exception of Islamic Albania) together with the overseas European populations of North America and Australasia experienced sustained, irreversible declines in fertility. Similarly, the timing of fertility change in developing countries has shown little relationship to the level of macro-economic development, despite the use of many different indicators. China, the southern states of India, central Java and Sri Lanka provide examples of poor, rural and largely agricultural populations in which fertility has fallen. Conversely, there are highly urbanized, and relatively affluent populations, notably several Arab States, where high fertility persists.

Contrary to the implication of most economic reasoning about fertility, modes of production also appear to be unrelated to fertility change. Low fertility is not incompatible with a continued dominance of family enterprises, nor does the advent of wage labour necessarily herald fertility decline. Rural Thailand is a good example of rapid fertility decline in a population where family-owned farms are prevalent. The high fertility copper-belt of Zambia is one of many settings, demonstrating that wage labour (dominant for several generations in the case of the copper-belt) may not lead to a reduction in fertility. The same absence of a relationship is found when individual families rather than populations are compared. Extensive analyses of data from the World Fertility Survey have shown that, within countries, the fertility of households whose head is a wage-earner is no different from that of families where the head is self-employed or works for a relative (Rodriguez & Cleland, 1981).

Expectations drawn from economic theories about the impact of female employment on the opportunity costs of childbearing and hence on the level of fertility itself have also been rebutted by empirical evidence (United Nations, 1987). Neither at the societal level nor at the individual level has it been possible to establish a clear-cut relationship. Differences in fertility or level of contraceptive use between women who work and those who do not are small and inconsistent in direction.

The last major theme concerns the value of children as a form of security in old age and more generally as mitigators of risk. This idea was part of some of the earliest formulations of demographic transition theory, but has been developed and applied with particular force to Bangladesh by Mead Cain (Cain, 1978; 1983). He maintains that uncertainty due to environmental hazards is compounded by a degree of lawlessness that permits the extortion of land from the weak by the powerful. Land, therefore, is no guarantee of security, and other institutional supports at times of crises, including lateral kinship bonds, are weak. Thus the only dependable form of security is provided by the vertical lineage and in particular by mature sons. The best survival strategy for Bangladeshis, Cain maintains, is high fertility.

This is an apparently convincing analysis and yet, even for Bangladesh, it rests on the slender empirical foundation of a single village study and some implausible assumptions. Inherent in the thesis of children as forms of insurance, is the assumption of long time horizons. Until they are mature, children probably increase family vulnerability rather than decrease it. The insurance motive for childbearing therefore involves a very long term investment. But it is generally considered that agriculturists have short planning horizons (Foster, 1965).

A second implicit assumption is that parents actually anticipate the possibility of dependency on children. For men in rural India, at least one study suggests that this may not be so. Economic activity persisted until close to the time of death and there was no prolonged period of dependency due to senility or terminal illness (Vlassoff & Vlassoff, 1984). For widows, the dependence on sons appeared to be emotional rather than economic.

A third assumption in the thesis that moderate to large numbers of children, particularly sons, may be needed for security in old age and at times of adversity, is a lack of alternatives. Cain may be right to assert there are no alternatives in Bangladesh, but in most other societies it seems more likely that the kinship support system is broad based and does not rest upon an exclusive reliance on children. These considerations may account for the failure so far to find a link between security provision by the state and fertility behaviour (e.g. Nugent, 1985). Surprisingly, the topic has not been thoroughly researched and no definitive conclusions can be reached. But in the present state of knowledge, the persistence of high fertility as an adaptation to the economic insecurity of parents is no more than an hypothesis.

Taken separately, the negative evidence concerning these four themes or expectations derived from economic theories of fertility does not assume great significance. Monocausal explanations of changes in reproductive behaviour have never been convincing. Taken together, however, the evidence constitutes a fundamentally damaging critique of the view that the economic transformation of societies – involving the demise of the family as the unit of production, the growth of non-familial economic roles for women, and the advent of alternative non-familial forms of security – provides a satisfying explanation for fertility transition in Europe or more recent trends in developing countries. At the very least, other influences must be at work that distort or blur the demographic impact of economic modernization. It is to these other factors that attention now turns.

Positive evidence for the importance of cultural factors

Perhaps the strongest evidence for the importance of culture as a fertility determinant is the remarkable way in which the transition to lower fertility occurs within broad regional or cultural blocks. As already mentioned, the onset of fertility decline in Europe and its overseas settlements occurred within a relatively short span of time despite huge inequalities between countries in their economic attainments. Within a historical perspective, the same broadly synchronous movement of fertility holds true for Latin America, the Sino-influenced states of East Asia and the Caribbean. With a few exceptions, the Arab world and sub-Saharan Africa have yet to start the transition to lower fertility.

The same imprint of cultural affiliation on reproductive behaviour is evident within countries. Study of the European fertility transition has revealed numerous examples of linguistic, regional and religious differences in fertility trends that defy economic explanations. In Portugal, for instance, fertility fell in the less developed south long before it did in the more prosperous, but more conservative north (Livi-Bacci, 1971). In Belgium, there was a striking difference in the timing of fertility transition between the Flemish and Walloon communities (Lesthaeghe, 1977). The Soviet Union exhibits the largest and most persistent inter-community differences in fertility. Despite fifty years of structural change including the collectivization of agriculture, universal education, and mass employment of women, fertility remains high in the predominantly Islamic republics. For women born in 1946–50, the mean number of children born by age 40 ranged from below replacement level in the western and Baltic republics to over four births per woman in Uzbekistan, Tadzhikistan, and Turkistan (Blum, 1988).

Similar linguistic, ethnic and religious variations in fertility are common in developing countries. Moreover, extensive survey data have permitted rigorous statistical analysis, demonstrating that economic and educational factors cannot account for these cultural variations. For instance, contraceptive practice remains lower and fertility remains higher among Malays than among Chinese and Indian communities in Malaysia, even after controls for standard of living, rural–urban residence and education (Peng & Abdurahman, 1981). Similarly the high fertility of Moslim minorities in Sri Lanka and Thailand persists after the introduction of appropriate socio-economic controls.

The speed of fertility decline within culturally homogeneous populations and its rapid spread to all socio-economic strata is much more consistent with explanations that place an emphasis on the spread of knowledge, new ideas or aspirations than those that stress the paramount importance of changes in economic structure. The latter changed too slowly to account for the speed of fertility decline. Declines in fertility of 30% or more over a decade are not uncommon and the quick diffusion of birth control practices from educated, urban elites to rural populations has been documented in many countries by means of repeated surveys. The agent of change in these instances, surely, must be something that all families can experience, regardless of the huge diversity in their economic circumstances.

A persistent finding of multivariate analyses of contraceptive use and fertility is that parental education has a greater influence than economic characteristics. Fertility is invariably lower in families where either parent has secondary schooling. The experience of primary schooling usually leads to greater attempts to regulate fertility, but the effect of birth control on childbearing may be offset by curtailed breastfeeding. The net result is that fertility in some countries, particularly in Asia and Africa, may be higher among couples with a modest exposure to formal schooling than among couples with no schooling (Cleland & Rodriguez, 1988). Aggregate analyses parallel these family specific patterns. The level of adult literacy is a much stronger predictor of low fertility than income, urbanization or energy consumption (Cutwright, 1983).

There are many possible interpretations of the ubiquitous education–fertility relationship, but it is entirely consistent with the view that receptivity to new knowledge and ideas is a key determinant of fertility change. At this juncture it is worth pointing out that education and cultural identification exert a strong influence on childhood mortality: an influence that appears to be independent of, and equal in magnitude to, material standards of living. The schooling of the mother is of particular importance, but the precise links have yet to be established. Similarly, adult literacy is a stronger correlate of life expectancy at the national level than per capita income or other economic indicators. It is thus probable that the forces behind fertility and mortality decline are similar and that both involve changes in knowledge and outlook.

To conclude, evidence of the importance of cultural factors on the timing of fertility decline is irresistibly strong. Two further questions immediately arise from this conclusion. Do cultural variations in fertility behaviour reflect differences in the value attached to children (i.e. demand for children) or differences in attitude towards birth control? And what specific cultural factors lie behind the bland cultural labels of ethnicity and region to produce the observed differences in fertility? Attempts to answer these questions are made in the next two sections.

Motives or means

In addressing the issue of whether cultural variations in fertility reflect differences in demand for children (i.e. motive) or differences in attitude towards birth control (i.e. means), or both, the credibility of various types of evidence becomes critical. In particular, how can demand or desire for children be measured?

The instinct of an economist is to regard fertility itself as a valid measure of demand. If the level of childbearing is high, then it must mean that large numbers of children are needed or desired. An alternative approach is to argue that the idea of any conscious desire or demand for children is meaningless in societies where there is little or no birth control. Until the advent of choice, in the form of socially approved, relatively harmless methods of regulating births, childbearing is largely a fatalistic acceptance of what God or Nature brings.

A third school of thought maintains that it is possible to assess desire or demand for children by verbal enquiry, even in non-contracepting societies, and that, at least in the short to medium term, there may be appreciable differences between such verbally expressed desires and fertility behaviour itself. The differences may be attributed to barriers of one sort or another that obstruct the expression of latent desires: these barriers that may take the form of moral or social sanctions against birth control or may concern more prosaic matters of inadequate knowledge or access to methods, fear of side-effects and so on.

I incline to the third viewpoint for the following reasons:

1. Lack of ability to control reproduction does not preclude the formation of latent preferences, even though they are likely to be held with low intensity and contain an element of rationalization. For instance, we have no hesitation in expressing our desires about weather, over which we have no control.
2. The idea of a preferred number of children does not appear to be alien even among uneducated men and women in traditional societies.
3. Survey measures of desired number of children, or whether any more children are wanted, are reasonably reliable when assessed by re-interview techniques.
4. Survey measures are broadly consistent with the results of more intensive anthropological enquiries.
5. Even in developed countries where birth control is well established, many conceptions are not only unplanned but unwelcome and many births unwanted. For instance, a large fraction of the decline in fertility in the United States in the last 20 years is attributable to the advent of superior birth control methods, that reduced the number of unwanted births from 20% to 7% (Pratt *et al.*, 1984). If this substantial divergence between desired and actual fertility was true, until recently, in a highly developed

country, there is no reason to doubt its existence in less developed countries.

If the broad validity of survey measures of desired fertility is accepted, what picture emerges of reproductive aspirations in developing countries over the past 30 years? And to what extent are there cultural variations in the desire for children?

In sub-Saharan Africa, very high levels of desired fertility have been recorded consistently, with mean desired family sizes of six to eight children, and few respondents in surveys reporting a wish to cease childbearing. By contrast, in most of Asia and Latin America, desired fertility, as measured in surveys over recent decades, has been of modest dimensions, typically in the range of two to four children (though conditioned in much of Asia by the desire for at least one son).

The contrast in fertility desires between sub-Saharan Africa and other developing regions demands an explanation. The reason does not lie in general economic development or level of mortality. Many African countries have higher living standards and lower mortality than, for instance, much of South Asia. Similarly, the low densities of population in most of Africa cannot be the crucial underlying factor because many Latin American countries also have low population densities without such a high demand for children. The labour intensity of African agriculture, based on the hoe rather than the plough, has been proposed as a key distinction. But, surely, rice growing is equally demanding of labour inputs but is not associated with high fertility. Indeed, the main rice growing areas of Asia (e.g. South India, China, Thailand, Indonesia) are now characterized by rather low fertility. In my view, a more plausible, but still speculative, reason for the distinctiveness of African sentiments regarding children lies in social organization. The relative absence of strong nation-states in Africa to guarantee physical security may have been conducive to continual local conflicts that conferred an advantage on groupings with a large numerical size. This feature may have engendered strong cultural supports for high fertility.

Some commentators (e.g. Frank & McNicoll, 1987, Frank, 1989) identify the cause of African pronatalism in the institution of the family. Because of weak conjugal bonds, accentuated by polygyny, men do not shoulder the main costs of childbearing, which becomes the mother's responsibility. Yet children belong to the male lineage and claims can be made on them in later life. Thus men can gain the long term advantage of large numbers of children without the penalty of initial expenditures. Consequently, they find high fertility advantageous. Women, having no independent assets or land, have little choice but to fulfil male expectations about numbers of children. Others have seen the root causes of high African fertility aspirations in religious terms (Caldwell & Caldwell, 1987). In traditional African religious belief, ancestors retain significance and identity as active members of the lineage after death as long as they are specifically remembered and may be reborn in the form of children named after them. There is thus a premium on having a large number of living descendants because they enhance survival of ancestors and probability of rebirth.

Whatever the nature of the forces that underpin high demand for children in Africa, it remains to be seen how resistant they are to social change. The immutability of pronatalist attitudes should not be exaggerated. Zimbabwe, Botswana and Kenya already provide examples of fertility decline. The policies of African governments are undergoing a radical shift towards recognition of the need to reduce birth rates and this is probably the precursor of similar changes among the general population.

With the major exception of sub-Saharan Africa, cultural variations in demand for children tend to be modest, certainly far less than variations in fertility itself (Lightbourne, 1987). In other words, the contrasts between, for instance, low fertility Java and high fertility Pakistan, or between the low fertility Sinhalese and the high fertility Moors in Sri Lanka do not appear to stem primarily from differences in the number of children that parents say that they want. This finding suggests that the propensity to act upon fertility desires is related to cultural factors and that this variation explains why some cultures respond very quickly to economic change by fertility reduction while others do not.

Culture and fertility regulation

The major conclusions of the preceding section were that the persistence of high fertility in sub-Saharan Africa reflects a high demand for children, but that, elsewhere in the developing world, cultural variations in fertility do not appear to originate in differences in family size desires. This latter conclusion raises the probability that resistance to the means of fertility regulation may be culturally determined and that this feature may be the key to an understanding of the diversity of fertility trends of the last 30 years. In this section, the possible links between culture and birth control attitudes are discussed in more specific terms. The aim is to go some way towards identifying the characteristics that may determine these attitudes. Three factors merit particular consideration: religion, openness to outside knowledge and ideas; and gender relationships.

With regard to the first factor, it is clear that all religions to some extent, but in particular Christianity and Islam, offer potentially powerful sources of resistance to the idea of birth control. In both the Bible and the Koran, children are portrayed as a divine blessing, and exhortations to multiply abound. By implication, birth control may be interpreted as an interference with divine will. The explicit opposition to most forms of birth control by the Roman Catholic church is well known and it is only quite recently that the legitimacy of church rulings on the subject has collapsed, with the result that the Catholic–non-Catholic fertility gap has largely disappeared in North America and Europe (Mosher & Goldscheider, 1984). It is more easily forgotten that similar opposition by many more Christian sects existed in the nineteenth century. Indeed, birth control was one of the most controversial subjects of the day.

While effective Christian opposition to birth control has gone, the same may not be true for Islam. To be sure, many Islamic authorities have pronounced in favour of contraception (though not induced abortion nor, in many instances, sterilization), and an increasing number of governments of predominantly Islamic countries have started to promote family planning. Yet at local level, opposition by religious leaders may remain. This suggestion cannot be backed by direct empirical evidence, but it seems to me to be no coincidence that the main area of high fertility, outside sub-Saharan Africa, comprises the Islamic heartland, or that Muslim populations in multi-cultural states nearly always have higher fertility than other populations.

Closely related to the theme of religious opposition to birth control is the suggestion that cultures vary in the openness to outside knowledge and ideas and that this characteristic conditions the speed of acceptance of birth control. This thesis rests on the twin assumptions that birth control is a genuinely new practice for traditional societies and that it is perceived as an innovation from outside. The balance of evidence suggests that both assumptions are valid.

We know rather little about population control in pre-modern societies. Even in such well documented civilizations as classical Greece and Rome, where concern at over-population was widely expressed, it is unclear whether coitus interruptus was practised, or the extent to which induced abortions took place (Wilkinson, 1978). We do know that fertility in all human societies has been kept well below the biological maximum (thought to be about fifteen births per woman on average) by a range of restrictions on marriage and mating and by prolonged breastfeeding. But the mechanisms of population control in these societies often took the form of social rules whose anti-natal effect was not their conscious purpose, or which offered no individual choice, as for instance was the case with the systematic infanticide of twins, or of children born at inauspicious times.

In Europe, the routine regulation of births within marriage appears to have been absent and coitus interruptus, the main mechanism of the European fertility transition, came as a genuine behavioural innovation (Knodel and Van de Walle, 1986). Similarly, surveys in developing countries show that contraception was rarely practised until recently. In more isolated countries, Nepal for instance and much of tropical Africa, even knowledge of any method of contraception was sparse. It seems reasonable to conclude, therefore, that contraception is as much of an innovation in developing regions of the world as it was in Europe. The fact that recently developed methods (sterilization, intra-uterine devices and hormonal contraception) account for the majority of users in developing countries strengthens their conclusion. No doubt, knowledge of and resort to abortion have been more common historically but, until recent times, this method has either been unreliable or hazardous and thus probably used only at times of utmost necessity.

To the extent that birth control is perceived as a largely western innovation, it is probable that its acceptance will be closely linked to the degree of resistance to western values and ideas in general. Again, it is the Islamic world that has been able to withstand most effectively the onslaught of western secular values, and it is in these countries that the legitimization of birth control has been slow, relative to the stage of economic development.

The last cultural trait to be discussed is gender relationships. It is commonly argued that sexual inequality represents a major barrier to fertility decline. One implicit assumption underlying this argument is that women are instinctively more favourable than men to the idea of family size limitation, because they bear the main burden of childbirth and upbringing. But, in countries where women are granted little autonomy or status, their inclinations are subordinated to male opinions.

In fact, the empirical record shows little evidence of divergence between the reproductive aspirations of husbands and wives. Though most survey research on fertility and family planning is directed towards women only, the attitudes of both sexes have been investigated in a diversity of settings. In these instances, desired family sizes of men and women are typically found to be similar. Nor have the results suggested that men are particularly hostile to family planning. Furthermore, as was mentioned earlier, there is no close relationship between female employment outside the home (surely a key indicator of women's autonomy) and fertility behaviour.

It is thus easy to exaggerate the importance of gender inequality as a fertility determinant; but in at least one regard it does represent a barrier to the spread of contraception. The seclusion of women within their homesteads, as is the case in North India and many Islamic countries, greatly reduces their access to birth control information and services. Most modern contraceptive methods are designed for use by women and it is therefore easy for husbands to regard the whole topic as something that concerns their wives more than them. In this context, the confinement of women to their own homes and immediate neighbourhood can act as a severe barrier to contraceptive adoption. In Bangladesh, for instance, a visit by a woman to a clinic, for health or family planning advice, requires the prior consent of the husband and accompaniment on the trip by another member of the family. Thus what in other cultures is a trivial undertaking becomes in countries like Bangladesh a major feat of persuasion and logistics. In view of the effort required of women to seek contraceptive advice, combined with ambivalence and fear, it is perhaps not surprising that contraception has been slow to spread in countries with a tradition of female seclusion.

The role of government interventions

No discussion of fertility determinants in developing countries is complete without a consideration of the influence of government policies and services. Of the topics investigated by demographers in the last 25 years, few have generated such a vast literature but such little consensus as the impact of family planning programmes on fertility. One reason for this diversity of opinion is differences in the theoretical and political positions of those who write on the subject. In the more starkly economic theories of fertility transition, the influence of government policies tends to be dismissed on a priori grounds. The level of fertility is seen as the result of parental demand or need for children and there is little that government propaganda can do to alter the situation. The mere provision of birth control services cannot make a major impact, because couples can always control their fertility (by abstinence or withdrawal, for instance) if they are determined to do so. As should be clear from the preceding discussion, my own view is very different. If the legitimacy of birth control is a determinant of the timing of fertility decline, then government action to popularize and promote family planning may be a catalyst for change in reproductive behaviour. The key issue, of course, is how far governments or other formal institutions can run ahead of popular sentiments.

A further reason for divergence of views stems from differing definitions of programme impact. Many evaluations take as their starting point the number of couples who use contraceptives provided by a government programme and, from this figure, calculate the number of births prevented, on the assumption that couples would use no method in the absence of the programme. This calculation is termed gross effect and its underlying methodology is now fairly well established (United Nations 1979, 1986).

By the criterion of gross effect, many government family planning programmes are successful. Both survey and other estimates demonstrate that large proportions of users obtain supplies from government rather than private or commercial outlets. For instance, among 23 developing countries with an official programme and relevant survey data from a nationally representative sample, an average of 63% of users of modern methods in the rural sectors obtained supplies from a government source; the corresponding figure for the urban sector was 52% (Population Information Program, 1985). Many of these countries have high overall levels of contraceptive use and the number of averted births attributable to government services is very large.

Use of services, and resulting estimates of the gross demographic impact, is an essential first step in judging the merit of any family planning programme. However, this approach cannot answer the more fundamental issue of net effect, which takes into account the likelihood that, in the absence of a government family planning programme, couples may find other ways to control their fertility, either by obtaining modern contraceptives through the private sector or by employing traditional means such as withdrawal and abortion. The issue of net effect is thus addressed to the following question. To what extent would the course of fertility have differed had there been no official sponsorship of birth control? As this is a hypothetical question, it cannot be answered with scientific certainty, nor is it readily amenable to statistical analysis. Nevertheless, a partial answer can be provided with reasonable confidence.

Let us start with a proposition to which there can be no dissent. Family planning programmes, however defined, are not necessary for the practice of birth control to spread and marital fertility to decline to very low levels. This was true for Europe and remains true for the developing world. Guyana, for instance, has experienced a major fertility decline despite pronatalist governments; similar declines have occurred in some Latin American countries (e.g. Paraguay and Venezuela) in a context of government indifference; and fertility has probably fallen in Burma, despite government discouragement.

It is also the case that most government sponsored programmes have been introduced *after* appreciable marital fertility decline has already taken place. In this regard, governments have been followers rather than leaders, reluctant to commit resources and reputation before the need for family planning services had been clearly demonstrated by their citizens. The role of non-government organizations in identifying and responding to need in the early stages of fertility transition has often been crucial in the evolution of government policies.

The first empirical issue is whether official sponsorship of family planning can accelerate a decline in fertility that is already underway. One influential body of evidence comes from the work of Berelson, Lapham and Mauldin (e.g. Mauldin & Berelson, 1978; Lapham & Mauldin, 1984). Their major contribution has been to derive objective measures of the strength of national family planning programmes, thereby giving explicit recognition to the fact that many governments have official population or family planning policies but make little effort to implement them. Their scale of strength of effort includes items on the political and legal framework; the nature of services offered including adequacy of training, administration and supervision; the accessibility of particular methods; and record keeping and evaluation activities. In cross-national analysis of birth rate declines and the level of contraception, these researchers, and most others who have used the same data sets, have found that the strength of family planning programmes correlates with fertility decline and contraceptive use, after controlling for level of development. Thus, among countries at the same level of socio-economic development, those with vigorous programmes have experienced greater declines in the crude birth rate and record higher levels of use than those with weak programmes. Not surprisingly, socio-economic development and programme effort are closely intertwined. Of the 93 countries included in the most recent study, there is a strong correlation between level of development and the programme effort score, and evidence that programmes are more effective when overall development is high. These relationships reflect the fact, mentioned earlier, that spontaneous fertility decline, itself partly an outcome of social development, has typically preceded and stimulated government provision of family planning services. It also reflects the fact that poorly developed countries usually lack the trained staff, and communication and logistical infrastructure, to create services that would rate highly on the scale of effort developed by Lapham & Mauldin.

The most important role of government action in settings where spontaneous decline pre-dated any serious government programme has been to accelerate the diffusion of birth control from urban and educated sectors to less privileged sectors of society. Evidence to support this view comes from analyses which have shown socio-economic differentials in fertility and contraceptive use to be greater in countries with weak programmes than where strong programmes exist (Entwistle *et al.*, 1986). The speed with which contraceptive use has spread to all sectors of society in countries such as Thailand and the Republic of Korea, and the concomitant rapid decline in fertility can be partly attributed to the hand of government.

The achievements of government policies in countries where there was no prior decline in marital fertility and little prior use of contraception is very different. These are typically programmes launched by governments who have been convinced of the need for family planning by the macro-economic arguments for reducing population growth (arguments often pressed strongly by donor organizations), rather than by the expressed demand of citizens for services. As intimated earlier, ther are relatively few such countries who have made more than a nominal commitment to the provision of birth control services before the spontaneous advent of birth control. Among the better known examples are India, Bangladesh, Pakistan and Indonesia in Asia, and Mexico and Tunisia. What conclusions can be drawn from these diverse countries?

Mexico is perhaps the maverick among this group. Until the early 1970s, this country was also an anomaly within Latin America. Despite a rather high level of development, fertility changed little, amidst declines in most other countries of the region. In 1973, contraception was estimated to be practised by only 10% of couples. In this year, the government abruptly changed from a pro-natalist to an anti-natalist policy and launched a vigorous family planning programme. The level of contraception subsequently rose to 32% in late 1976 and reached 40% in 1978.

There are several possible interpretations of this sequence of events but the most plausible is that government action acted as a catalyst which released a huge latent demand for birth control and resulted in one of the most rapid transitions in reproductive behaviour ever recorded. The timing of government action was no doubt fortuitous, and there is also little doubt that contraception would have spread and fertility declined in the absence of government intervention. But it is most improbable that the change would have been so dramatically abrupt, had the government position remained unchanged.

In Tunisia, the case for attributing the decline in marital fertility to government intervention is stronger than in Mexico, because the cultural setting of the Arab region was clearly not conducive to the spontaneous spread of birth control. Family planning services and education were instigated in Tunisia in the late 1960s by President Bourguiba, amidst a range of secular, reformist measures, including an emphasis in female education. It is thus difficult and inappropriate to disentangle the direct effect of government sponsored family planning activities from the indirect effects of wider social reforms. However, it is incorrect to attribute the decline in fertility in this country and the relatively high level of contraception solely to educational advances. What distinguishes Tunisia from other Arab countries is the much higher level of use among women with no schooling. It takes excessive scepticism to deny the success of the Tunisian programme in encouraging fertility control among less privileged sectors of the community.

Tunisia and neighbouring Algeria provide a fascinating contrast. Algeria has in many ways a more developed economy but, in reaction to a bitter war of independence, pursued social policies that reinforced its Islamic traditions. There was for instance, little attempt to modify the position of women or, until very recently, to popularize family planning. Throughout the period 1960–80, fertility remained constantly high at over seven births per woman. In contrast, Tunisia, over the same period, experienced a decline in fertility of about 30% and reached a fertility level of about five births in 1980. This seems to me an obvious instance of the influence of government policies on the reproductive behaviour of citizens.

Consider now the experience of the four poor, largely agricultural countries of Asia, who all rate low on levels of socio-economic development. All have had strong centrally directed programmes, with considerable pressure on staff to reach monthly or annual targets (usually expressed in terms of number of family planning acceptors recruited) and financial compensation paid to acceptors of certain methods.

There is one case of obvious failure: the Pakistan programme of 1965–1969. This was a vigorous programme with ample funding and top-level political support which received international acclaim at the time. But the whole edifice collapsed with the fall of the Ayub regime and with the results of the 1968–1969 Impact Survey which revealed that the programme had had virtually no effect; only five per cent of married women reported use of contraception.

Many possible reasons for the failure have been proposed but, in retrospect, such an ambitious crash programme in a conservative and largely illiterate society, was probably doomed. Not surprisingly, subsequent governments have been reluctant to make a serious effort to promote family planning and the political climate under President Zia, with its partial return to basic Islamic values, was not conducive to such policies. Fertility in Pakistan certainly has not fallen and may even have risen slightly. In the mid 1980s contraception was practised by only about 10% of married couples. If government policies towards family planning had been more positive would fertility have declined? No-one can be sure, but it seems likely that Pakistan would now be in a similar position to India and Bangladesh, to which attention now turns.

In these two countries, levels of contraceptive use have risen slowly but steadily to over 30%. In India, there is unmistakable evidence of fertility decline, while in Bangladesh the evidence of decline is more recent and its magnitude is uncertain. In both countries it is almost certainly correct to attribute most of the modest change in reproductive behaviour directly to programme efforts. Few couples, outside major cities, seek contraceptive advice and supplies spontaneously or from private sources; most require considerable prior motivation from field staff and there is little doubt that the financial compensation offered by the government exerts some influence on decisions (Satia & Maru, 1986). In brief, the promotion of family planning in India and Bangladesh has proved to be a difficult and expensive undertaking, which, despite the sustained political commitment, has achieved only modest success so far.

In the case of Bangladesh, there is convincing evidence that the weak impact of the programme is not, as is so frequently assumed, an inevitable consequence of the very low level of development of the country, or of its social structure. Reformulation of field strategies and better supervision, it seems, can lead to substantial increases in contraceptive use and fertility declines. The experience in the experimental Matlab area and the more recent extension of the lessons learned to other areas of the country constitutes the strongest evidence that appropriate provision of contraceptive services and education can achieve a substantial demographic impact among one of the poorest and least educated populations in the world (Phillips *et al.*, 1988).

In Indonesia, large increases in contraceptive use are not confined, as in Bangladesh, to a few semi-experimental areas. In Java and Bali, which together account for 60% of the total population, contraceptive prevalence rose from about 2% in the early 1970s to about 40% in 1980. Furthermore, use is as high, if not higher, among the poorest and least educated strata of society as among more privileged strata. It can of course be argued that social and economic conditions in the early 1970s were particularly conductive to fertility decline, but the central influence of the government family planning programme on demographic trends is difficult to deny.

We may conclude from these examples – Mexico, Tunisia, Pakistan, Bangladesh, India and Indonesia – that government intervention has been able to initiate fertility transition, though the experience has been very mixed, encompassing total failure; modest impact achieved by costly and rather severe forms of action; and a large degree of success. Paradoxically, those programmes that have been most widely acclaimed for their success (e.g. Thailand and Colombia) may have altered the course of events less than programmes widely criticized for the modesty of their achievements (e.g. India and Bangladesh).

Conclusion

Explanations for fertility transition are dominated, both in formal theories and in popular thinking, by economic concepts and assumptions. The central driving force is reduced demand for children brought about by economic modernization. Within this framework, little attention is typically given to the means of regulating family size; rather they are assumed to be available once the need arises.

The purpose of this essay has been to demonstrate that economic change does not automatically translate into demographic change. Countries at the same level of economic development – as measured by labour force composition, per capita income or urbanization – may have very different patterns and levels of fertility. Thus the link between economic structure and family sizes is far weaker than is usually claimed.

This variable demographic response to modernization is clearly related to cultural boundaries. At one extreme is China and the other Sino-influenced states of East Asia, all conspicuous for very rapid transitions to low fertility. At the other extreme is the Islamic heart-land stretching from North Africa to Pakistan, where, despite relatively advanced economies in many instances, fertility typically remains high.

The explanation for these cultural variations in fertility does not appear to lie primarily in differences in the size of family that couples desire. Rather, the propensity to translate preferences into appropriate behaviour appears to be the more important discriminant. Perhaps many commentators, particularly from societies where birth control is a normal part of life, have underestimated the significance of the change from natural to regulated fertility, and underestimated potential resistance to it stemming from religious values and/or general antipathy towards western secular innovations and ideas.

Ultimately, of course, low levels of human fertility will be universal, for no society can for long sustain high rates of population growth. Ambivalence towards birth control will evaporate in the Islamic World just as it did in Europe. However, this longer term perspective in no way detracts from the important influence of cultural factors in the short to medium term.

References

Blum, A. (1988). Descendance atteinte dans les generations Sovietique et dans ses republiques. *Population*, **43**, 1143–51.

Cain, M. T. (1978). 'The household life cycle and economic mobility in rural Bangladesh'. *Population and Development Review*, **4**, 421–38.

Cain, M. T. (1983). Fertility as an adjustment to risk. *Population and Development Review*, **9**, 688–702.

Caldwell, J. C. & Caldwell, P. (1987). The cultural context of high fertility in sub-Saharan Africa. *Population and Development Review*, **13**, 409–37.

Cleland, J. & Rodriguez, G. (1988). The effect of parental education on marital fertility in developing countries. *Population Studies*, **42**, 419–42.

Cutright, P. (1983). The ingredients of recent fertility decline in developing countries. *International Family Planning Perspectives*, **9**, 101–9.

Entwistle, B., Mason, M. W. & Hermalin, A. J. (1986). The multi-level dependence of contraceptive use on socio-economic development and family planning program strength. *Demography*, **23**, 199–216.

Foster, G. M. (1965). Peasant society and the image of limited good. *American Anthropologist*, **67**, 293–315.

Frank, O. (1989). Societal constraints to effective family planning in sub-Saharan Africa. Paper presented at International Union for the Scientific Study of Population Seminar, Tunis, June 1989.

Frank, O. & McNicoll, G. (1987). An interpretation of fertility and population policy in Kenya. *Population and Development Review*, **13**, 209–43.

Knodel, J. & Van de Walle, F. (1986). Lessons from the past: policy implications of historical fertility studies. In *The Decline of Fertility in Europe*, ed. A. J. Coale & S. C. Watkins. Princeton: Princeton University Press.

Lapham, R. J. & Mauldin, W. P. (1984). Family planning programme effort and birth rate decline in developing countries. *International Family Planning Perspectives*, **10**, 109–18.

Lesthaeghe, R. J. (1977). *The Decline of Belgian Fertility 1800–1970*. Princeton: Princeton University Press.

Lightbourne, R. E. (1987). Reproductive preferences and behaviour. In *The World Fertility survey: an assessment*, ed. J. Cleland & C. Scott, pp. 838–61. Oxford: Oxford University Press.

Livi-Bacci, M. (1971). *A Century of Portuguese Fertility*. Princeton: Princeton University Press.

Mauldin, W. P. & Berelson, B. (1978). Conditions of fertility decline in developing countries, 1965–1975. *Studies in Family Planning*, **9**, 89.

Mosher, W. D. & Goldscheider, C. (1984). Contraceptive patterns of religious and racial groups in the United States 1955–1976: Convergence and distinctiveness. *Studies on Family Planning*, **15**, 101–11.

Nugent, J. B. (1985). 'The old-age security motive for fertility'. *Population and Development Review*, **11**, 75–97.

Peng, T. N. & Abdurahman, I. (1981). Factors affecting contraceptive use in Peninsular Malaysia *WFS Scientific Reports 23*. Voorburg, Netherlands: International Statistical Institute.

Phillips, J. F., Simmons, R., Koenig, M. A. & Chakraborty, J. (1988). Determinants of reproductive change in a traditional society: evidence from Matlab, Bangladesh. *Studies in Family Planning*, **19**, 313–34.

Population Information Program (1985). Fertility and Family Planning Programs: an Update. *Population Reports*, **Series M**, 8.

Pratt, W. F., Mosher, W. D., Bachrach, C. A. & Horn, M. C. (1984). Understanding U.S. fertility: findings from the National Survey of Family Growth Cycle III. *Population Bulletin*, **39**, 5.

Rodriguez, G. & Cleland, J. (1981). Socio-economic determinants of marital fertility in twenty developing countries: a multivariate analysis. In *World Fertility Survey Conference 1980: Record of Proceedings 2*, pp. 337–414. Voorburg, Netherlands: International Statistical Institute.

Satia, J. K. & Maru, R. M. (1986). Incentives and disincentives in the India family welfare program. *Studies in Family Planning*, **17**, 136–45.

United Nations (1979). *Manual IX. The Methodology of Measuring the Impact of Family Planning Programme on Fertility*. New York; Department of International Economics and Social Affairs.

United Nations (1986). *Addendum to Manual IX*. New York: Department of International Economic and Social Affairs.

United Nations (1987). *Fertility Behaviour in the context of Development: Evidence from the World Fertility Survey*. New York: Department of International Economic and Social Affairs.

Vlassoff, M. & Vlassoff, C. (1980). 'Old age security and the utility of children in rural India'. *Population Studies*, **34**, 1, 487–99.

Wilkinson, L. P. (1978). Classical approaches to population and family planning. *Population and Development Review*, **4**, 439–55.

11 *Understanding recent fertility trends in the Third World*

ALLAN G. HILL

Introduction

The full force of social controls on the speed of human reproduction can be clearly appreciated when we contrast reproductive potential with realized levels of fertility. Although the calculation of potential fertility, or fecundity, is beset with conceptual, technical and measurement problems, a consensus has emerged that for most human populations, the upper limit on the average number of live births per woman must lie between 13 and 17 (Bongaarts & Potter, 1983: p. 79). This is not to deny that some individual women can and do produce many more than this number, but for populations rather than individuals in which biology is the only factor restricting fertility, an acceptable average number of births per woman is around 15.3. Even this number of births can only be achieved by a rare combination of circumstances. We can list these very briefly:

- continuous exposure to the risk of conception between menopause and menarche, meaning a steady sexual relationship throughout the woman's fertile life;
- complete avoidance of any contraceptive method, including 'natural' methods, during the entire reproductive lifespan;
- no use of induced abortion or of any activity which might be undertaken to provoke a miscarriage;
- avoidance of any breastfeeding practice.

Clearly, this combination of conditions is very rare, helping to explain why measured levels of 'natural'[1] marital fertility rarely surpass 11 or 12 live births. Some historical data based on written records or family reconstitution methods have thrown up a few populations in which most of the above conditions are met (see the summary table in Leridon, 1977: pp. 107–9). Today, examples from the developing countries are quite rare since very few traditional contemporary societies eschew totally the use of breastfeeding even though the prevalence of contraceptive use may be quite low. For a variety of purposes, including the estimation of the number of births averted by family planning programmes, it is nonetheless worthwhile trying to estimate potential fertility and the relative importance of each of the main factors determining the level of realized fertility. The value of this approach is that it highlights the contribution of primarily biological factors in the determination of human fertility, as well as indicating some of the complexities involved in the interaction between biology and behaviour.

The proximate determinants schema

Following Bongaarts & Potter (1983), we can write:

$$\text{TFR} = \text{PF} \times C_m \times C_c \times C_a \times C_i$$

where TFR = the estimated value of the total fertility ratio;
PF = the level of potential fertility (fecundity);
C_m = the index describing the effect of marriage on overall fertility;
C_c and C_a = the effect of contraception and induced abortion respectively on the level of natural fertility;
C_i = the effect of breastfeeding on the duration of postpartum amenorrhoea.

Each of the indices is constructed in such a way that the range of possible values extends from 1.0, meaning that the particular proximate determinant has no effect at all on the reduction of fertility below the level of potential fertility, to values close to 0.0, implying a very strong reduction in potential fertility. In other words, subtracting the value of any of the indices from one is a measure of the proportionate reduction in potential fertility attributable to that proximate determinant, holding all other variables constant.

The overall accuracy of the model as a tool for summarizing the main factors restricting potential fertility can be evaluated by comparing the value of total fertility estimated through the model with that measured directly. Generally, the association is very strong, as illustrated by Figure 11.1, but as Casterline *et al.* (1984) have noted, there are many problems with the details of the assumptions incorporated in the model itself, as well as with the accuracy of reporting of sensitive aspects of fertility such as use of contraception and induced abortion. An analysis of the proximate determinants of fertility by educational group and by residence reveals many inconsistencies. One of the centrally important findings made by Casterline *et al.* (1984) is that the indices of the proximate determinants tend to be associated, despite the assumption of independence. The plot of the index of marriage against the index of contraception for educational sub-groups in 29 countries shows that cross-sectionally at least, rising levels of education are positively associated both with later marriage and with

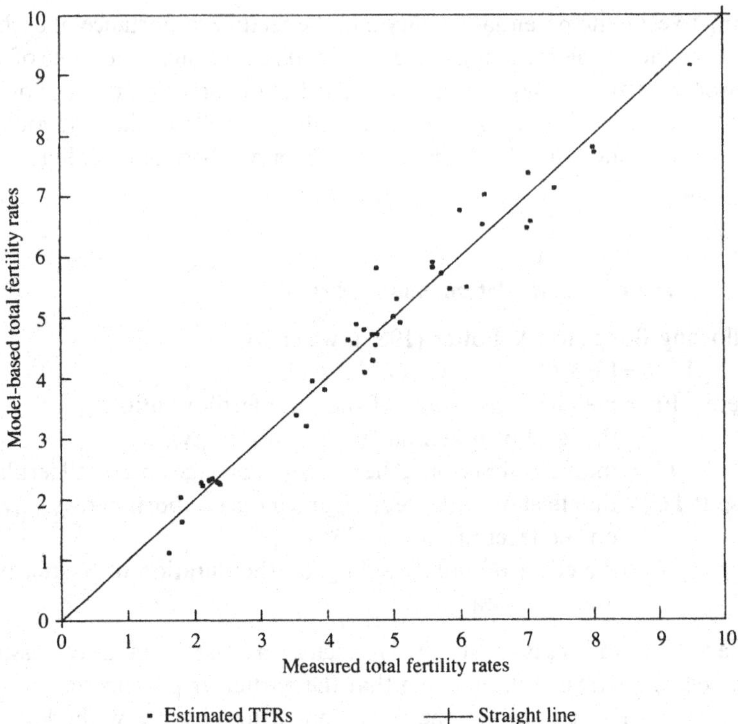

Figure 11.1. Measured and model-based total fertility rates compared for 41 developed and developing countries. *Source*: Plotted from data in Tables 4.3 and 4.4 in Bongaarts & Potter (1983).

more widespread use of contraception (Figure 11.2). Figure 11.3 illustrates for the same educational sub-groups the strong negative association between later marriage and duration of postpartum amenorrhoea which, as discussed later, is largely determined by the duration and intensity of breastfeeding. Finally, on Figure 11.4, we see illustrated the key inverse relationship between the index of contraception and the index of postpartum amenorrhoea. It is this trade-off between more widespread use of contraception and the declining duration of breastfeeding as educational status rises which is at the heart of the problem of interpreting fertility trends and prospects in poor countries. In regression terms, even after controlling for the independent effects of marriage, one unit change in the index of postpartum amenorrhoea is offset by roughly equal changes in the index of contraception. The strength of the relationships illustrated in Figures 11.2 and 11.3 and 11.4 indicates that there are common cultural or socio-economic factors which simultaneously affect the main behavioural factors affecting fertility, but that changes in the proximate determinants

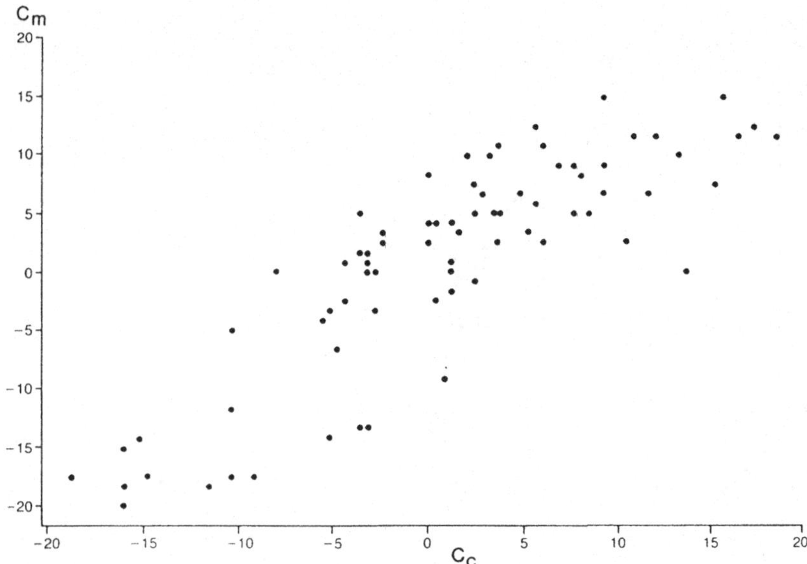

Figure 11.2. Deviations from the country values for the indices of marriage (C_m) and of contraception (C_c) for educational sub-groups in 29 countries. *Source*: Casterline *et al.* (1984).

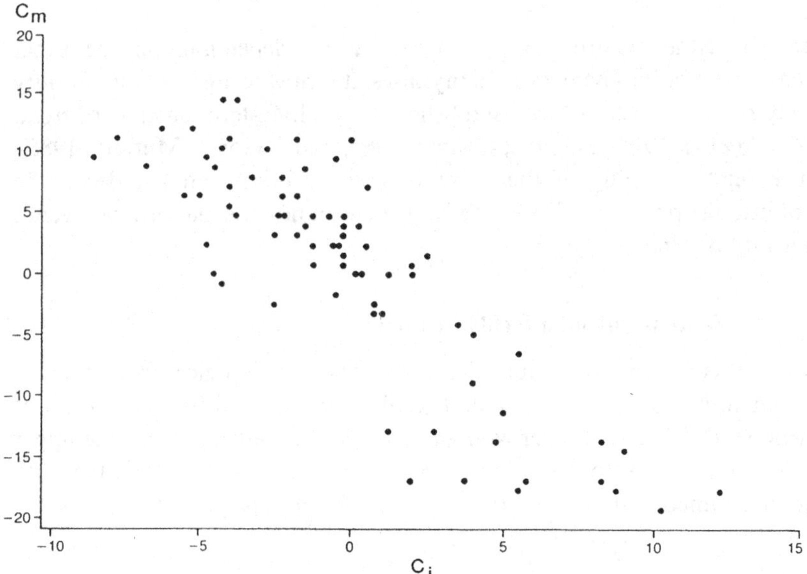

Figure 11.3. Deviations from the country values for the indices of marriage (C_m) and of postpartum infecundability (C_i) for educational sub-groups in 29 countries. *Source*: Casterline *et al.* (1984).

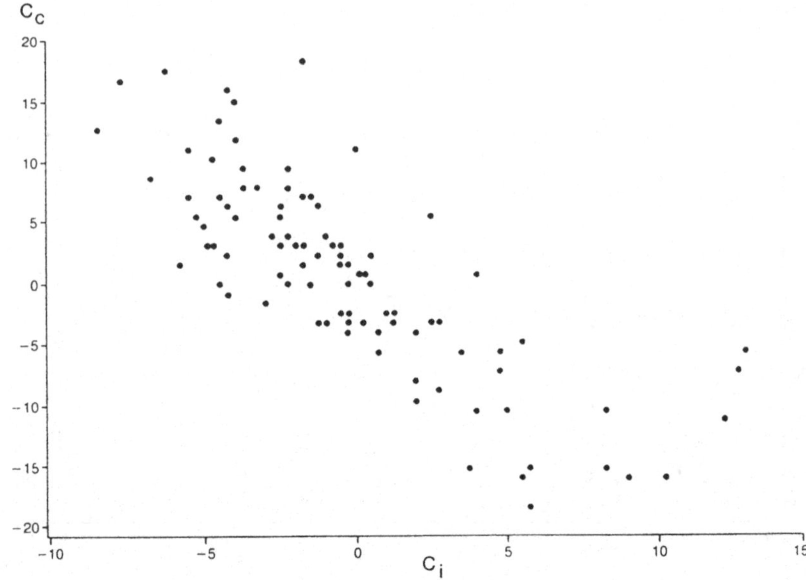

Figure 11.4. Deviations from the country values for the indices of contraception (C_c) and of postpartum infecundability (C_i) for educational sub-groups in 29 countries. *Source*: Casterline *et al.* (1984).

can be either reinforcing or compensating, depending on the exact magnitude of the changes. In many cases, it is now recognized that fertility may rise in advance of the establishment of a long-term downward trend (Coale *et al*, 1979 and 1982; Collver, 1965, and Dyson & Murphy, 1985). The relative timing of the onset of contraception and the decline in traditional patterns of breastfeeding determines the pattern of overall fertility decline.

Why is potential fertility so low?

First, it is worth setting out briefly the principal biological factors which result in an estimate of potential fertility or the total fecundity ratio at around 15.3 live births per woman. Broadly, biology alone sets the upper and lower bounds to the scale of human reproduction but the exact fertility level attained within these limits involves both biology and behaviour.

Sterility

In every human population, there will always be some couples who can never produce a live birth as well as some who will never achieve a recognizable conception. The proportion in these two categories is difficult to measure when contraception is widespread, but in natural fertility populations, primary sterility can be as low as 3% overall. The proportions sterile rise steeply above age 30, as shown both by historical data (Leridon, 1977) and by the results from the World Fertility Survey (WFS) discussed in detail by Vaessen (1984). Levels of primary sterility vary widely in some Third World countries because some diseases directly affect fertility and also because of particular customs such as circumcision and poor obstetric practice. Sterility rises with age both because of biological changes related to the reproductive system, and because of complications related to previous pregnancies. In the WFS, definitions of sterility were based both on behaviour (no live births in a period of five years for women living in a sexual union and not using contraception) and on direct questions ('As far as you know, is it physically possible for you and your husband to have a child, supposing you wanted one?'). The importance of the association measured by the WFS between the demographic definition of sterility (i.e., no births for five years amongst exposed women) and the direct question about the couple's potential ability to bear a child is that respondents are clearly conscious of their infertility or low fecundity. This will affect sexual behaviour and may in some cases affect one or more of the other proximate determinants of fertility. Udjo (1985) has measured exceptionally short periods of lactation amongst the Kanuri of north-eastern Nigeria amongst whom sterility is known to reach exceptional levels. Couples with proven fecundity are also more likely to use a more reliable method of contraception, thus affecting our method-specific estimates of contraception effectiveness (see the discussion in United Nations, 1985 and in Westoff, 1974 for more details).

Whilst the measurement of sterility appears to be a relatively simple task, there are many complications related to the definition of a recognizable conception and a live birth, as well as factors such as the selection of fertile women into marriage. In societies in which fertility is highly prized and in which sexual activity and pregnancy are acceptable before marriage, the sterile or even those with of doubtful fecundability[2], may be avoided as marriage partners or be prone to frequent divorce. The surveys in Mali and by Udjo amongst the Kanuri provide support for this behaviour, which may further increase the proportions childless simply because those with low fecundability may not be given the same chance as others to demonstrate their ability to bear children. Thus, social customs related to the choice of marriage partners may be re-enforcing the much stronger genetic tendency for the genes of fertile couples to be strongly represented in the next generation. In any case, the existence of some sterile women in every population reduces the number of births per woman in the population as a whole. The rapid rise in the proportions sterile over age 30 is important when some marriages occur at high ages.

Fecundability

Although fecundity in a general way can be defined as the inverse of sterility, the more precise term, 'fecundability', is helpful in understanding the relative importance of biological and behavioural elements in the determination of the monthly probability of noting a conception. Four separate stages in the process resulting in a recognizable conception can be identified and each has a set of attendant risks and uncertainties.

(i) Anovulatory cycles

Leridon (1977: p. 14) cites several studies which suggest that between 5 and 15% of all menstrual cycles are anovulatory, but these overall figures have little meaning without some consideration of age. The age-specific data cited by Gray in Leridon & Menken (1979: p. 225) show that amongst 12–14 year olds, as many as 90% of cycles may be 'abnormal', whereas in the late 20s, over 80% of cycles are 'normal' (i.e., ovulation occurs and the luteal phase is of normal length). Bongaarts & Potter (1983: p. 32) put the proportion of anovulatory cycles during the central childbearing years as 5% and accordingly use 0.95 as the average probability of ovulation occurring in a menstrual cycle.

(ii) Intercourse during the fertile period

The period when insemination can result in fertilization is surprisingly short – perhaps 48 hours in all. Fertilization can take place if insemination occurs a short time before or after ovulation, as sperm retains its fertility for 24 to 48 hours and the ovum remains viable for 12 to 24 hours after release from the ovary. The likelihood of conception is thus strongly affected by the frequency of coitus within a fertile cycle, assuming that the timing of coitus is independent of the timing of ovulation. A simple formula allows the calculation of fecundability given any specified level of coital frequency. Although data on coital frequency are rare for developing countries, there is some evidence to suggest that overall frequencies may be a little lower than those in developed countries (Leridon, 1977: table 3.5). Privacy and contraceptive availability are just two external factors which influence frequency and regularity of intercourse.

(iii) Incomplete fertilization

Fertilization does not always occur when sperm and ovum meet. There may be genetic problems as well as difficulties with the implantation of the fertilized ovum on the wall of the uterus. Although empirical data on these risks are difficult to collect, Bongaarts & Potter (1983: p. 33) have proposed that we accept an estimate of 0.95 as the probability of successful fertilization following the meeting of sperm and egg.

(iv) Recognizable conception

For most women, the first sign of pregnancy is a missed period. Between fertilization and the successful establishment of the fertilized ovum in the uterus, losses can occur. Again, Bongaarts & Potter (1983: p. 33) suggest that we work with an estimate of 50% for these losses until better data become available. Leridon, using data from clinical sources, re-calculated the life table probability of observing a live birth from a population of 100 ova exposed to fertilization. From the first part of this life table, he estimates that 42 of these ova would be continuing pregnancies at the second week following ovulation and that these ova would later result in 31 live births (Leridon, 1977: p. 81).

Taking all these risks together, we can see that the probability of having a recognizable conception in a fertile cycle is the joint probability of the four independent factors outlined above. We know from empirical data that fecundability (f) ranges from 0.2 to 0.3. From the above, the probability of having an ovulatory cycle (p_1) is estimated at 0.95; the probability of intercourse taking place during the fertile period (p_2) varies with coital frequency; the probability of successful fertilization (p_3) is about 0.95; and the probability of having a recognizable conception (p_4) is about 0.5. Therefore, it is possible to estimate fecundability and the waiting time to conceive for different frequencies of intercourse. Coital rates of between 7 and 11 acts per menstrual cycle would produce estimates of fecundability covering the range from 0.2 to 0.3, with corresponding waiting times to conceive of between 7 and 5 months. These estimates are in line with the empirical data on frequency of coitus by age and with model and empirical estimates of fecundability.

Intra-uterine mortality (miscarriages)

There are numerous technical and practical issues which make the measurement of spontaneous foetal loss very difficult. Intra-uterine death rates amongst pregnancies in progress at four weeks from the last menstrual period fall rapidly with gestational age. Perhaps 8% are lost in the first four to seven weeks, and less than 1% after the 20th week (Leridon, 1977: pp. 80–1). Rates vary widely by age of woman, being twice as high amongst women over age 40 and probably a little higher amongst teenagers.

This brief summary of the risks associated with different phases of the reproductive process serves to illustrate some of the quite general biological restrictions on potential fertility amongst all human populations. Two points are noteworthy. First, very few of the reproductive processes outlined above appear to be very strongly influenced in a direct way by nutritional status. Clearly, in populations which are starving, many physiological as well as behavioural changes related to reproduction will occur. In mildly malnourished populations, however, the waiting time to conceive, the reciprocal of fecundability, does not vary significantly by the nutritional status of the mothers (Bongaarts & Delgado, 1979; Bongaarts, 1980). Second, whilst sterility and the foetal loss rate are both higher in populations in which pelvic inflammatory diseases and the sexually transmitted diseases are common, there appear to be few other direct links between health and fecundity. There are, however, many indirect links mediated by behavioural changes which will be described later.

Breastfeeding and postpartum amenorrhoea

The link between frequency and intensity of suckling and the suppression of the ovarian function is now well established, although many of the biological details remain unclear (McNeilly *et al.* 1985). The duration of amenorrhoea in populations of women can be predicted quite accurately on the basis of knowledge about the duration and pattern of breastfeeding. This knowledge helps to explain the wide diversity of fertility levels in populations with natural fertility, as well as in the contemporary Third World. Historical data from a French village in Quercy indicate that women married between 1700 and 1791 had a completed family size of just 3.7, while amongst French Canadians marrying at age 20 in the early eighteenth century or Hutterites in the USA marrying at age 20 in the period 1921 to 1930, completed family sizes were as large as 11 children (Leridon, 1977: table 7.1). In addition, knowledge of the nature of the link between breastfeeding and the length of the postpartum non-susceptible period helps to interpret some of the puzzling recent changes in fertility in the Third World. Both the collapse of traditionally long and intense patterns of breastfeeding and the disappearance of other traditional practices such as abstention from sexual relations for some months or years following the birth of a child have shortened birth intervals in non-contracepting populations in Africa and the Middle East. The reasons for abandonment of breastfeeding are not always because bottle-feeding is considered better or easier. Other considerations are important, including the demand by husbands for earlier resumption of sexual relations in societies such as those in West Africa in which traditional polygynous marriage is becoming much harder to maintain. The complex web linking these changes with earlier weaning and child fostering has been documented by Bledsoe (1987) and Bledsoe & Isiugo-Abanihe (1989). Some notable work by Lunn (1985) has shown how maternal nutritional status can influence breastfeeding patterns and hence birth intervals in a surprising and indirect way. Both examples serve to underline the 'accidental' way in which fertility and fecundity can be influenced by changes not directly linked with the reproductive process itself. Indeed, one important conclusion to be drawn from the foregoing section on reproduction in natural fertility populations is that the level of fertility is almost a by-product of the patterns of marriage and infant feeding, as well as involving the social customs regulating relations (including sexual) between the sexes.

Contraception and abortion

In any discussion of Third World fertility, it is impossible to omit all consideration of contraception and induced abortion even though their effects on fertility would seem rather obvious. Contraceptive prevalence levels are rising steadily and in some parts of the Third World, declared induced abortion rates are substantial (see the data in Population Reports, 1980). All the published abortion rates are probably too low even where such practices are legal. One piece of evidence for this is that the potential fecundity rates calculated by dividing the total fertility rates by the product of the indices of the other proximate determinants of fertility are mostly low in comparison with model estimates of potential fecundity. The underestimation could stem from errors of measurement or from omission of some effects such as those attributable to undeclared use of induced abortion. By making some reasonable assumptions about the expected level and pattern of induced abortion in developing countries, Casterline *et al.* (1984) were able to adjust the model-based estimates of potential fertility from WFS data and produce revised estimates more in line with expectations. In some countries such as Guyana, Haiti and amongst the better educated women of Pakistan, the adjustments are substantial, indicating undeclared total induced abortion rates of at least one child per woman (Casterline *et al.*, 1984: table 21).

This increase in the use of contraception and induced abortion does not necessarily result in reduced fertility. First, there are off-setting effects between contraceptive use and breastfeeding practice summarized in Figure 4. This inverse relationship helps to explain why in countries such as Zimbabwe, rising rates of contraceptive prevalence (38% in 1984) are not immediately associated with rapidly falling fertility. The total fertility ratio in Zimbabwe in 1984 was 6.5, only one child less than the completed family size of the women aged 45–49 when interviewed (Bongaarts, 1987: p. 137). Second, an induced abortion only averts about 0.4 of a birth in non-contracepting populations (slightly more when contraceptive prevalence rises) since some pregnancies would in any case result in a miscarriage and the period of infecundability following an abortion is shorter than that after a live birth.

Applications to contemporary societies

It is possible to find today non-contracepting populations in the Third World in which fertility is either surprisingly high or surprisingly low. In Africa, for example, the WFS data for rural Kenya indicate a marital total fertility ratio of 9.5 live births per woman whereas amongst the Dobe !Kung of the Kalahari, marital total fertility rates are below 5 births per woman (Mosley *et al.*, 1982; Howell, 1979). Here, we will present data on several contrasting fertility regimes in the Third World and discuss the possible fertility consequences of the ways in which societies move from natural to controlled fertility.

Fertility in the West African Sahel

In a series of surveys begun in 1983, information on fertility and its main proximate determinants was collected on three ethnic groups in Mali–Bambara millet farmers, Fulani semi-nomads, and nomadic Tamasheq herders (see Hill, 1985 for full details). Total fertility rates ranged from 5.2 to 8.1 births in five clearly non-contracepting populations, all resident in the same general area (Table 11.1). The highest fertility rates were found amongst the settled Bambara millet farmers and the lowest amongst the Kel Tamasheq groups interviewed in the large bend of the Niger known as the Gourma. There are many details of the social and economic organization of these different groups which are relevant to the interpretation of their fertility patterns but here we will only consider the summary of the proximate determinants of fertility and the indices summarized in Table 11.1. Very briefly, whilst there are some minor differences between the five groups in the duration of breastfeeding and consequently, the duration of postpartum amenorrhoea, the major differences in fertility levels are largely due to contrasting marriage patterns.

Table 11.1. *Fertility and its proximate determinants in five populations in central Mali*

	Population				
		Fulani		Tamasheq	
	Bambara	Masina	Seno	Masina	Gourma
% of women currently married	90	81	84	73	64
Mean duration of breastfeeding (months)	20	18	23	18	21
Mean duration of amenorrhoea (months)	13	12	16	12	14
Total fertility ratio	8.1	7.1	6.6	6.6	5.2
Index of marriage (C_m)	0.92	0.86	0.88	0.67	0.68
Total marital fertility ratio	8.8	8.3	7.5	9.8	7.6
Index of lactational amenorrhoea (C_l)	0.63	0.65	0.58	0.65	0.61
Index of spousal[a] separation (C_s)	0.95	0.95	0.95	0.96	0.93
Potential fertility	14.7	13.4	13.6	15.8	13.6

[a] This index measures the fertility-reducing effects of the temporary separation of husbands and wives due to migration.
Source: Hill (1985: p. 61).

Practising the levirate and widespread polygyny, the Bambara have developed an efficient system for marrying girls early and keeping women in marriage despite the vagaries of widowhood and divorce. Overall, 90% of women of reproductive age were married at the time of the surveys. This contrasts sharply with the situation amongst the Kel Tamasheq amongst whom less than three-quarters of women of reproductive age were married. These figures are exceptional as far as Africa is concerned and compare with the very low proportions married in western Europe in the nineteenth century, including Ireland after the Great Famine (Drake, 1963). Again, the explanation of this distinctive Tamasheq marriage pattern is complex and well explained in two recent theses (Randall, 1984; Winter, 1984) but the crux of the matter appears to be the Tamasheq descent and inheritance system which, at least for the noble classes, allows women to own cattle and other property in their own right. This allows them to retain some degree of economic independence and hence social independence, allowing them to resist some of the pressure to marry early and to remain married. This system also includes a strongly enforced taboo on pre-marital sexual relations amongst nobles.

These data illustrate some of the difficulties of describing patterns of fertility as 'traditional' or 'modern'. They also show very clearly that the role of pure biology is to limit the possible range of fertility levels which can be achieved in non-contracepting populations. Fixing the exact level of achieved fertility within this range appears to be a matter of economics and social organization rather than of physiology.

Fertility in the Arab countries

Comparison of the fertility patterns in the Sahel with those of selected Arab countries illustrates very clearly the point that different fertility levels can be achieved by quite different combinations of the proximate determinants. The Malian groups achieve total fertility ratios of between 5.2 and 8.1 by a combination of some celibacy, restriction of most sexual activity to within marriage, and a number of practices including prolonged breastfeeding and in some cases, avoidance of sexual relations for a significant period following a birth. Contraception, traditional or modern, does not figure as an important variable, despite every effort to uncover such practices during the surveys. Sterility in the Malian populations does not appear to attain abnormal levels.

By contrast, the factors behind the determination of fertility levels in the Arab countries are quite different. Of the four Arab populations shown in Table 11.2, the pattern of the proximate determinants in the Yemen Arab Republic is the closest to the African examples shown on Table 11.1. The notable difference is the much shorter period of breastfeeding and consequently a curtailed period of postpartum amenorrhoea (seven months) in Yemen, compared with figures closer to one year in Mali. In Tunisia, Jordan and Syria, despite the relatively high rates of contraceptive prevalence (32%, 23% and 18% respectively), fertility was high because of the relatively short duration of breastfeeding. Birth intervals in the Arab countries have dropped to close to two years (Aoun, 1989). Apart from the decline in breastfeeding, there appear to be few physiological restraints on fertility. Long periods of postpartum abstinence are not a feature of most Middle Eastern societies, and fecundity appears to be relatively high in the Arab countries (see Appendix C in Casterline *et al.*, 1984).

Some of the most valuable insights into the determinants of fertility come from an analysis of educational and residential differentials within a single country. Some data for Jordan in 1976 are presented in Table 11.3. The extraordinarily high fertility of women who have never been to school or who live in rural areas is remarkable. Parities as high as this do not appear in the 1961 census data for the East Bank, suggesting that fertility has indeed risen in recent years. For both the uneducated and the rural women

Table 11.2. *Indices of the proximate determinants of fertility in selected Arab countries*

		Total fertility ratio	C_m	C_c	C_i
Jordan	(1976)	7.6	0.75	0.80	0.81
Syria	(1978)	7.5	0.71	0.82	0.80
Tunisia	(1978)	5.8	0.65	0.75	0.74
Yemen A.R.	(1979)	8.4	0.86	0.98	0.76

Sources: Casterline *et al.* (1984) and Aoun (1989).

Table 11.3. *Differentials in the indices of the proximate determinants of fertility for Jordan (1976)*

	Total fertility ratio	C_m	C_c	C_i
1. Education (years)				
None	9.3	0.85	0.91	0.77
1–3	8.7	0.88	0.79	0.78
4–6	7.0	0.78	0.72	0.82
7+	4.8	0.60	0.60	0.60
2. Residence				
Rural	9.4	0.84	0.93	0.77
Other urban	7.7	0.74	0.76	0.81
Major urban	6.3	0.69	0.70	0.84

Calculated from: Singh *et al.* (1985, tables 1 and 2).

in Jordan, the mean durations of breastfeeding and hence of postpartum amenorrhoea are both relatively short, particularly in comparison with the mean durations for the Malian groups. The index of the effect of postpartum amenorrhoea on fertility is therefore much closer to one in Jordan than for all five of the Malian groups shown on Table 11.1. Amongst the women in major urban areas and particularly amongst women with seven or more years of schooling, contraceptive use is common and exerts a strong influence on fertility. It is in the intermediate categories of education and residence where we see most clearly the offsetting effects of relatively short breastfeeding and modest use of contraception. Taken together, marriage, contraception and breastfeeding reduce potential fertility amongst the women with one to three years of education by 46%. The figure is just slightly greater, 54%, for women with four to six years of education. The interaction of these factors explains why the kind of fertility transition under way in the Arab world is more like Type I on the schematic diagram summarizing three possible routes to lower fertility (Figure 11.5).

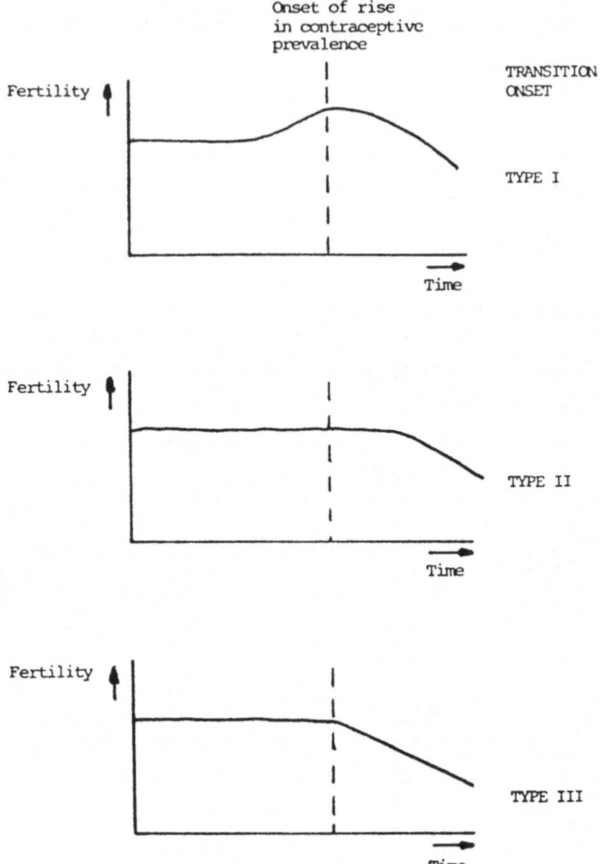

Figure 11.5. Schematic representation of three ways in which fertility changes occur in relation to the onset of widespread use of contraception. In Type I situations, natural fertility rises in advance of contraceptive use. In Type II cases, early advances in contraceptive use are offset by rising natural fertility. In Type III, contraception becomes widespread without any major rise in natural fertility. *Source*: Bongaarts (1986).

Conclusion

The process of moving from high to low levels of fertility was first described in terms of a simple model known as 'the demographic transition'. The central assumption was that improving mortality rates, particularly for children, would lead to increased rates of natural increase in the short term but, subsequently, pressures arising from this increased rate of growth would make themselves felt at both the societal level and at the level of individual households. These pressures would then provoke a fall in fertility. This is a caricature of an idea which now has many additional dimensions (Caldwell, 1976). The basic idea has been called into question by a careful re-analysis of the European data on which the model was supposedly based (see Coale & Watkins, 1986 for a summary). There are now many competing theories to explain fertility trends and differentials in the Third World (see Simmons, 1988 for a recent review) but very few are flexible enough to accommodate the complex set of biological and behavioural relationships summarized above. In effect, the admittedly cross-sectional but nonetheless international analysis of the changing values of the indices of the proximate determinants of fertility summarized in Figures 11.2, 11.3 and 11.4, suggests that several forms of the fertility transition are likely. Some possible fertility outcomes relative to the onset of family planning programmes have been developed (see Figure 11.5) but even this is a gross over-simplification of what might occur, because similar levels of fertility or similar rates of change can be achieved by quite different combinations of the proximate determinants. This is the conclusion from the comparison of the West African and Arab examples above.

The important point to be drawn from the study of these contrasting cases is that for high fertility populations in Africa and the Middle East to approach the fertility regimes of Latin America and of south-east Asia, major adjustments are involved, together with associated risks. Replacing traditional birth spacing practices with modern ones, or subtle arrangements for marriage and inheritance with rules decreed by the modern state, involves social change on a quite fundamental level. It seems that the smooth transition to lower fertility can only be accomplished in those populations in which there are social institutions and forces which allow several changes to occur simultaneously. In societies which are relatively closed to external influences and which have distinctive and strong indigenous systems of social organization, these simultaneous changes may not be possible.

Notes

1 'Natural' fertility is a term used to describe the pattern of fertility in the absence of contraception and induced abortion. It was first introduced by Henry (1953) with reference to historical populations.

2 Fecundability is usually defined as the probability of noting a conception during a fertile menstrual cycle. A high value would be 0.3; a typical value often used in reproductive models is 0.2.

References

Aoun, S. (1989). A comparative study of fertility change in Tunisia, Syria and the Yemen Arab Republic. Unpublished PhD thesis, London School of Hygiene and Tropical Medicine.

Bledsoe, C. H. (1987). Side-stepping the post-partum sex taboo: Mende cultural perceptions of tinned milk in Sierra Leone. In *The Cultural Roots of African Fertility*, Obafemi Awolowo University and the University of Pennsylvania.

Bledsoe, C. H. & Isiugo-Abanihe, U. C. (1989). Strategies of child fosterage among Mende grannies in Sierra Leone. In *Reproduction and Social Organization in Sub-Saharan Africa*, ed. R. Lesthaeghe. Berkeley: University of California Press.

Bongaarts, J. (1980). Does malnutrition affect fecundity? A summary of evidence. *Science*, **208**, 565–9.

Bongaarts, J. (1986). The transition in reproductive behaviour in the Third World. *Working Papers* No. 125, Center for Policy Studies. New York: The Population Council.

Bongaarts, J. (1987). The proximate determinants of exceptionally high fertility. *Population and Development Review*, **13**, 1, 133–40.

Bongaarts, J. & Delgado, H. (1979). Effects of nutritional status on fertility in rural Guatemala. In *Natural Fertility*, ed. H. Leridon & J. Menken. Liège: Ordina.

Bongaarts, J. & Potter, R. G. (1983). *Fertility, Biology and Behaviour*. New York and London: Academic Press.

Caldwell, J. C. (1976). Towards the re-statement of demographic transition theory. *Population and Development Review*, **2**, 3, 321–66.

Casterline, J. B., Singh, S., Cleland, J. & Ashurst, H. (1984). The proximate determinants of fertility. *WFS Comparative Studies* no. 39. Voorburg: International Statistical Institute.

Coale, A. J., Anderson, B. & Harm, E. (1979). *Human Fertility in Russia since the 19th. Century*. Princeton: Princeton University Press.

Coale, A. J., Cho, Lee-Jay and Goldman, N. (1982). Nuptiality and fertility in the Republic of Korea. In *Nuptiality and Fertility*, ed. L. T. Ruzicka. Liège: Ordina.

Coale, A. J. & Watkins, S. C. (1986). *The Decline of Fertility in Europe*. Princeton: Princeton University Press.

Collver, O. A. (1965). *Birth Rates in Latin America: New Estimates of Historical Trends and Fluctuations*. Institute of International Studies, Research Series no. 10. Berkeley, California.

Drake, M. (1963). Marriage and population growth in Ireland. *Economic History Review*, second series, **xvi**, 301–13.

Dyson, T. & Murphy, M. (1985). The onset of fertility transition. *Population and Development Review*, **11**, 3, 399–440.

Henry, L. (1953). Fondements théoretiques des mesures de la fecondité naturelle. *Revue de l'Institut International de Statistiques*, **21**, 3, 135–51.

Hill, A. G. (Ed.) (1985). *Population, Health and Nutrition in the Sahel*. London: Kegan Paul International.

Howell, N. (1979). *Demography of the Dobe !Kung*. New York and London: Academic Press.

Leridon, H. (1977). *Human Fertility*. Chicago: University of Chicago Press.

Leridon, H. & Menken, J. (eds.) (1979). *Natural Fertility*. Liège: Ordina.

Lunn, P. G. (1985). Maternal nutrition and lactational infertility: the baby in the driving seat. In *Maternal Nutrition and Lactational Infertility*, ed. J. Dobbing, Nestlé Nutritional Workshop Series No. 9. New York: Raven Press.

McNeilly, A. S., Glasier, A. & Howie, P. W. (1985). Endocrine control of lactational infertility. In *Maternal Nutrition and Lactational Infertility*, ed. J. Dobbing, Nestlé Nutrition Workshop Series No. 9. New York: Raven Press.

Mosley, W. H., Werner, L. H. & Becker, S. (1982). The dynamics of birth spacing and marital fertility in Kenya. *WFS Scientific Reports*, No. 30, Voorburg: International Statistical Institute.

Population Reports (1980). *Complications of Abortion in Developing Countries*. Population Information Program, Series F, No. 7, Johns Hopkins University.

Randall, S. C. (1984). A comparative demographic study of three Sahelian populations. Unpublished PhD thesis, London School of Hygiene and Tropical Medicine.

Simmons, O. G. (1988). *Perspectives on Development and Population Growth in the Third World*. Plenum Press, New York and London.

Singh, S., Casterline, J. B. & Cleland, J. G. (1985). The proximate determinants of fertility: sub-national variations. *Population Studies*, **39**, 1, 113–35.

Udjo, O. E. (1985). Levels and differentials in fertility and mortality among some Kanuri of north-east Nigeria. Unpublished PhD thesis, London School of Hygiene and Tropical Medicine.

United Nations (1985). *Studies to Enhance the Evaluation of Family Planning Programmes*. Population Studies No. 87, New York: Dept. of International Economic and Social Affairs.

Vaessen, M. (1984). Childlessness and infecundity. *WFS Comparative Studies*, no. 31. Voorburg: International Statistical Institute.

Westoff, C. F. (1974). Coital frequency and contraception. *Family Planning Perspectives*, **6**, 136–41.

Winter, M. (1984). A study of family kinship relations in a pastoral Toureg group of northern Mali. Unpublished PhD thesis, University of Cambridge.

12 *Monogamy, landed property and demographic regimes in pre-industrial Europe: regional contrasts and temporal stabilities*

RICHARD M. SMITH

In any consideration of possible relationships between monogamous marriage, resource exploitation and fertility in the European past, generalizations are hard to make with confidence. However, Europe provides an interesting milieu within which to review certain issues to do with these inter-connections because it is the continent which supposedly fostered the emergence of a marital regime that is believed to have been highly distinctive, perhaps unique, within a wider world context. Although Malthus had noted this possibility with regard to European society somewhat hesitantly and rather belatedly in his own writings, it was left to John Hajnal (1965) in a classic essay published a quarter of a century ago to specify these unique features with geographical, if not chronological precision. Europe's most distinctive feature with regard to marital behaviour lay in the fact that women married, as E. A. Wrigley (1985: p. 3) has succinctly noted, 'little and late, by comparison with other major cultures whose characteristics are known – India, China, Muslim countries or the Roman World ...' In these last-mentioned societies marriage for women began at, or close to, the arrival of sexual maturity and it was to a considerable degree regarded as improper both for the girl and her family if she were not married at or soon after puberty. In such circumstances marriage was pre-determined by a biological process or largely invariant moment in the life-course and was in no sense a 'moveable feast'. What is more, only a handful of women remained permanently unmarried; only those who suffered from severe mental or physical handicap would be likely to do so. In eighteenth-century Russia – a society most definitely not part of Europe in the past if we employ marital patterns as a geographical demarcator – early censuses, by their methods of recording the very few women over the age of 19 listed as never-married (fewer than four per cent), leave the historian with no grounds for speculating on the reasons for their spinster condition. In almost all cases such persons were designated to have been seriously handicapped; 'hunchback', 'halfwit', 'blind' are the telling marginal entries against the names of unmarried females in the manuscript census listings (Czap, 1978; Hoch, 1986: pp. 76–7). Women who succeeded in reaching the age of 20 without marrying were very unusual – indeed veritable oddities.

In Western Europe, women were not 'catapulted into marriage by a biological trigger' (Wrigley, 1985: p. 4), but married, delayed marriage or avoided it altogether in response to pressures which appear to have been largely social and economic. It is, of course, important not to undervalue the fact that these pressures were mediated through a myriad of regional customs and rituals. Women married late, at ages usually between 24 and 27 on average, and a substantial minority of women never married at all. The celibate proportion usually ranged from between 5 and 15 per cent, though occasionally that proportion might exceed 20 per cent, such as in late seventeenth-century England (Weir, 1984; Schofield, 1985) or late nineteenth-century Ireland (Fitzpatrick, 1985).

In a more recently published essay whose 'classic' status is only marginally less than that to which we have already referred, Hajnal (1982) emphasized two other features of what he termed the 'northwest European household formation system'. These were that marriage was, for the broad mass of the population, neolocal in that bride and groom set up their new residence apart from that of their respective natal families – often in a different community. A second and highly important feature of this marital residence pattern was that the new couple should be totally, or very largely, economically independent of their kin. Indeed this latter feature is a key element in the debate about the economics of household formation in that area of Europe in the past. In the contrast that Hajnal draws between the household formation systems in north-west European areas and those in many other parts of the world, he emphasized the greater costs incurred by couples in the former area in the establishment of their households. Couples forming independent households would experience specific 'start-up' costs that were not imposed upon young brides and their grooms who co-resided with their kin in joint households.

To acquire the resources needed to sustain an independent household young adults might have to spend a long time accumulating savings. Hajnal (1982: pp. 470–6) noted that in north-west Europe the interval between puberty and eventual marriage was passed by both males and females as servants in the households of those to whom they were generally unrelated and from whom they received payment in the form of bed and board and an annual stipend that was frequently retained by the employer until the service contract was completed. Hajnal believes that this institution gave flexibility to marriage timing since in difficult years or periods the service 'net' expanded to incorporate many who would otherwise have married earlier.

Hajnal (1965: pp. 133–4) had noted that resources necessary for the establishment of households by the newly married could be obtained through inheritance of family wealth, primarily, although not invariably, land. He realized that this could come about by waiting for a *post mortem* inheritance but it could also be facilitated by *pre mortem* redistributions of assets within the family.

Demographic and family historians have expended a great deal of energy in evaluating the impact of the assumed relationships between patterns of devolution or inheritance and male ages at marriage (Berkner & Mendels, 1978). While it would be cavalier to claim that the results achieved through this form of analysis have been of limited value, it is important to emphasize that considerations of inheritance patterns which are thought to relate primarily to male marriage age do not readily explain how womens' ages at marriage were determined. To the extent that any patterns are consistently detected in north-west Europe, labourers (the landless) and their wives appear to have married later than other groups (Derouet, 1980; Drake, 1969: chapter 5; 1974: pp. 71–5). For instance, a systematic study of a sample of German villages in the eighteenth and nineteenth centuries reveals that wives of labourers are characterized by later-than-average marriage age in virtually all of the communities (Knodel, 1988; pp. 130–5). Wives of farmers were distinguished by their earlier age at first marriage than wives of men in other groups. What is noteworthy is that men's ages at first marriage seem to have been less closely linked to their occupational category than that of their wives. Such evidence rests uncomfortably with the view that proletarianization led to the groups so effected marrying earlier because they were free from restraints imposed by property inheritance as a determinant of their household formation and because as wage labourers they were likely to achieve maximum earning at a young age. If the basic norm that a marital union should only be entered into when the prospects for being able to sustain a family seemed reasonable was more or less followed by all strata of society, persons in less favourable economic positions might have needed to delay marriage longer in order to accumulate sufficient savings, broadly defined in terms of the goods necessary to maintain a household. Since the wife brings such savings with her to the marriage, proletarian men may have opted for women who were older, given that such women would have had a longer time to acquire these savings (Sundt, 1980). Couples composed of husbands and wives from more advantaged social backgrounds would have been under less restraint in these respects, especially if the wife was provided with a relatively sizeable dowry and the husband was assured of inheriting sufficient wealth or a relatively secure source of livelihood, such as a family farm or a craft workshop or business. Where a bride had a dowry she might have been as marriageable at 20 as at 28, but where she depended upon her own efforts for the resources she brought to the marriage, she might have been in a much better position to attract an eager suitor at a later age.

The uniqueness of the European pattern lies primarily in the high age at marriage of women (often with a relatively small difference between the ages of husband and wife) rather than an advanced age of 28 to 30 for the groom and a precocious age of 16 or 17 for the bride, as is commonly found in many non-European areas. Of course, this feature has attracted the attention of scholars who have interested themselves in the provisioning of females with property or the theoretical access to property which they might acquire through inheritance or *pre mortem* means (Bonfield, 1986). This issue stands at the heart of a great many of the writings of that remarkably prolific, historically-minded social anthropologist, Jack Goody.

Goody, in a series of articles, beginning in the 1960s (1969, 1971, 1973a,b; Goody, Irving and Tahany 1971; Goody and Tambiah 1973; Goody and Buckley 1973) and culminating in an important monograph (1976), has been concerned with what he regards as a fundamental problem in comparative historical sociology. He wishes to provide a means of explaining the very real contrasts between Eurasia on the one hand and sub-Saharan Africa on the other. In pursuing this course of thinking, which is endowed with a distinctive intellectual coherence, he has provided a highly original account of what he sees as the uniqueness of European marriage and family patterns. Goody (1969) began his investigations by identifying two modes of inheritance which he termed respectively, 'homogeneous' inheritance and 'diverging devolution'. In homogeneous systems, found primarily in sub-Saharan Africa, inheritance is sex specific; men inherit from men and women from women. In Europe and Asia, on the contrary, women inherit from men (and vice versa), although there may be restrictions on the type and amount of property they can own or acquire in that way. The origins of the term 'diverging devolution' lies in the fact that while in homogeneous systems property is retained within the descent group on inheritance, in diverging systems the effect of the disposal of property to both sexes is to diffuse it outside the descent group.

Systems of diverging devolution are also those in which the direct vertical transmission of property from parents to their offspring is emphasized at the expense of inheritance across sibling groups. Goody sees the *pre mortem* acquisition of dowry, settled on a woman at marriage, as the major form of female inheritance in Eurasia. Dowry systems establish a conjugal fund, within which the properties of the husband and the wife are combined. Consequently, considerable care is taken to match the resources of each. Among other things, this fund ensures the support or endowment of a woman at and during her widowhood. In these circumstances Goody postulates that marriages would be strongly controlled by the parental generation, through the careful monitoring, indeed at times strict policing, of courtship. A fundamental ideology concerning pre-marital virginity is needed to sustain these controls. Although in theory in dowry systems, additional wives would bring additional resources, the matching of resources is thought difficult to duplicate, and there are problems in setting up several conjugal funds whether undertaken simultaneously or sequentially. Consequently, Goody (1976: p. 51) believes the effects of dowry and the conjugal fund make for widespread monogamy in Eurasia in contrast to Africa for 'in Africa . . . there are no such constraining factors to inhibit the tendency for men to accumulate women in sexual partnership.'

The scarcity of land in the Eurasian setting compared with its relative abundance in Africa is central to Goody's argument. The preoccupation with the maintenance of social status and the association of that status with land in limited supply links Goody's argument to the kinship system and the character of agricultural technology. He places great weight on the contrasts that follow from the respective roles played by the plough in Eurasia and the hoe in Africa. Goody (1976: p. 20) argues that the plough brings with it a major increase in agricultural productivity as it is one of the many inventions of advanced agriculture which 'permits an individual to produce much more than he can consume'. Intensification brings greater economic differentiation because it enables a ruling group to develop a much higher standard of living out of agricultural production, and because it allows the development of full-time specialists who do not produce their own food. At the same time the population expansion allowed by increased production exacerbates land scarcity and producers themselves become ranked on their command of the means of production. Differential holdings of land and capital emerge. These socially determined inequalities in wealth and the surpluses thereby created lead to numerous strategies to maintain possession of the land that underpinned such accumulations. These surpluses formed the basis of private and inheritable 'real' property and the laws and governments that guaranteed it. As Caldwell *et al.* (1989: p. 191) note in their endorsement of Goody, 'private farms of fixed size and defined boundaries are not easily compatible with large lineages whose members all demand access to common resources. They are better maintained and defended by families organized around a strong and usually indissoluble marriage bond.'

In contrast, African societies which traditionally lacked the plough and the wheel depended upon a plentiful supply of land of very poor quality on which productivity was low. Ploughs under these conditions were of limited value and over substantial tracts of territory the tsetse fly restricted the use of draft animals. This low productivity based on poor soils and simple technology produced little full-time specialization. While there were differences between rich and poor farmers, these tended, so Goody argues, to be based on the 'strength of one's arm or the number of sons'. Economic differentiation was minimal. The 'control of agricultural production really meant control over people, thus centering concern on reproduction rather than on land inheritance' (Caldwell *et al.*, 1989: p. 192).

Goody, in a controversial but brilliantly conceived sequel to his ideas on diverging devolution as they had developed from 1969, proceeds, in a book he published in 1983 entitled *The Development of the Family and Marriage in Europe*, to describe and to explain the process whereby European marriage and kinship patterns diverged from those they had once shared with a larger geographical area including North Africa and much of Asia. In the Eurasian system of diverging devolution Goody claims the tendency was for marriage to take place within the kin group. There is a simple economic explanation for the prevalence of in-marriage; by marrying their sons with their cousins german, the members of the elementary kinship group prevented part of the patrimony – the inheritance of the female – from leaving the clan. Indeed, he argues (1983: p. 33) that today 'close marriage continues to distinguish the Asiatic and African shores of the Mediterranean, running from the Bosphorus to the Maghreb, from the European one running from Turkey to Spain where the tendency is to marry exogamously'.

The change or break by which Europe severed itself from Asia and North Africa came with Christianization. More exactly, it began in the fourth century, at a time when a sect was transformed into a church, an organization whose interests required that it build up and defend a patrimony of landed wealth. The Church then constructed, in Goody's opinion, in an almost conspiratorial fashion, a system of rules which condemned existing practices which restricted inheritance of family goods to the direct male line. Goody therefore sees the Christian Church proscribing a set of actions, in particular adoption, polygyny, remarriage, divorce, concubinage and endogamy, that were to a very great extent concerned with what can be termed 'patriline repair'. Because of diverging devolution, these proscribed practices were strategically vital as repair of the highly vulnerable patriline was constantly called for. Demographers using various simulation tools have been able to show that under stationary demographic conditions (zero growth rate) in about 20% of instances a man would have no direct heirs of either sex alive to succeed him, and in a further 20% of cases he would have only daughters and no sons (Wrigley, 1978; Smith, 1984a).

Possessed of this degree of vulnerability, patrilines could be broken with relative ease by a Church that proceeded to emphasize family stability within a new framework which safeguarded the individual's liberty to dispose of his and especially *her* possessions. Such devolutionary potential favoured the mobility of land and its alienation and served to increase the probability that some land would gravitate towards the Church. For this reason, so Goody argues, the Church strove to break patrilines by encouraging marriage by mutual consent promoting full powers of testation and forbidding marriage within strictly defined degrees of kinship.

Population historians have yet to ponder the explicit and implicit links that Jack Goody makes between his treatment of ideology, canon law and practices, largely documented as relating to the uppermost echelons of society, and the demographic attributes and consequences of popular marriage in medieval and early modern Europe. Goody is explicitly arguing that any changes in behaviour that led to the social structural underpinnings of Hajnal's marriage pattern could not be directly associated with Protestantism or capitalism but must be located at least a millenium earlier in time. Implicitly, Goody is inclined to observe a degree of homogeneity – a common denominator – in the demography of popular marriage over those areas of Europe which came under the influence of the Christian Church and so set it apart from the Near East and North Africa. It must be assumed that he wishes his argument to apply to both Latin Christendom and the Eastern Church.

Indeed, one might suppose that certain demographic patterns would have become identifiable that coincided in fairly broad terms with the areas within the control of the Christian Church. Goody suggests that the increasing pressure to marry outside the kin group should have resulted in a delay in marriage in the adult life-cycle and an increase in the proportions eventually failing ever to marry. An emphasis too, on the conjugal bond and the promotion of the consensual basis of spouse choice might have served to bring about similar results through an extention of the courtship process. A reluctance to sanction, indeed a positive hostility towards remarriage should have increased the proportion of widows in the higher age-group in Christian as opposed to pre-Christian or Islamic populations. Furthermore, the campaign against concubinage or unsolemnized marriage, if it created a stronger 'legal' sense of what constituted an illegitimate birth, should have brought about a situation in which the onset of sexual relations would have begun only after the priest's blessing and the witness of the couple's union and the nuptial mass.

Alas, when these elements are investigated as an interrelated set of pan-European characteristics, whether we consider the more widely available demographic data of the early modern and modern periods, or the much sparser evidence for the centuries before 1500, we find little to suggest the existence of any uniformity in the demography of marriage in a Europe bounded on its southern edge by the northern shore of the Mediterranean sea.

We may reflect initially on the evidence collected by the Princeton European Fertility Project for the late nineteenth and twentieth centuries (Coale & Watkins, 1986). That project provides us with the first fully inclusive statistical cross-section of European nuptiality from data concentrated upon the year 1870 in the form of an index (I_m)[1]. In 1870, of all the component parts of fertility, nuptiality (I_m) was the most spatially variable. This feature is reflected in the fact that a coefficient of variation of I_m in 1870 was almost 21% (0.209) for the 579 provinces (excluding those of France where marital fertility had already begun to fall) compared with that for an index of marital fertility (I_g)[2] which was only 0.106 (Coale & Treadway, 1986: p. 48). Notwithstanding this variety, the data yielded by the European Fertility Project provided vivid confirmation of Hajnal's (1965) discovery of what he believed was a fault-line running between Leningrad (St. Petersburg) and Trieste. It was noted by Coale & Treadway (1986: p. 52) that the Leningrad–Trieste line did not serve as a wholly satisfactory divide between provinces with I_ms above and below 0.55 (a value thought to capture the threshold between late and low-intensity marriage on the one hand and early and high-intensity marriage on the other). Certain provinces to the west of the line, such as those in Southern Portugal, had I_m values above 0.55. In 31 of the 48 Spanish provinces in 1887 I_m values exceeded 0.60 and in 12 of the 16 Italian provinces in 1871 I_ms were over 0.55 – rising to values over 0.70 in parts of the extreme south. Indeed, these data in the more comprehensive form provided by the Fertility Project would seem to confirm the suspicion of Hajnal who was decidedly forthright about the differences between eastern and western Europe but cautiously noted (1965: p. 103) that 'significant departures from the European pattern may probably be found not only as one proceeds eastwards but on the southern edge of Europe as well. Parts of Italy and Spain are more like Greece than like Belgium or Sweden'.

What furthermore emerges from the analysis undertaken by Watkins (1986) of nuptiality patterns after 1870 is that the nation state constitutes a highly misleading unit within which to conduct investigations, since regions within countries displayed markedly different patterns and, with only a few exceptions, these relative regional differences were retained until the 1930s, even though the absolute levels of I_m shifted (in many cases downwards). What is more, the failure to find any consistent relationship between nuptiality and economic indicators, whether the analysis was synchronic or diachronic, led Watkins to argue strongly for regional intransigence in the matter of female marriage behaviour that, at least until a relatively late date, resisted any tendency towards convergence. There is even reason to suppose that at the level of small geographic units such as village communities, age patterns of marriage showed greater inter-village variation than did inter-occupational differences within villages (Knodel, 1988: p. 136).

The European Fertility Project, perhaps inevitably, given its brief, did not, at least in any depth, pursue the question of how durable these marriage geographies had been before 1870. It is important to note in relation to our discussion of Goody's thesis which is concerned with European-wide patterns and processes, that regional intransigence was stressed over and above temporal change. This is not an emphasis that has characterized much of the writing by historians and demographers on the medieval and early modern periods in Europe. A preoccupation with the nature and timing of 'transitions' has been very evident. The seeds for this view were sown in Hajnal's original study when he cautiously suggested that the 'European pattern' – that prevailing to the west of the line from Leningrad to Trieste – had come into existence between the later Middle Ages and the seventeenth century in England and possibly elsewhere in north-west Europe (Hajnal, 1965: p. 134). Hajnal based these views on a limited analysis of the results of J. C. Russell's (1948) investigation of certain late fourteenth-century Poll Tax returns which led him to conclude that the proportions of females over the age of 14 who were married were too high to be fully compatible with a European marriage pattern. Convinced that in much of north-west Europe a pattern of late female marriage was clearly present in the seventeenth century, Hajnal (1965: p. 132) in an uncharacteristically speculative passage suggested that a 'full explanation of the European marriage pattern would probably lead into such topics as the rise of capitalism and the protestant ethic'. Although these conclusions were never presented by Hajnal as firmly established, they fostered among certain scholars a tendency to see a European marriage pattern eventually emerging in association with sixteenth-century religious changes (Sharlin, 1977; Monter, 1979; Dupâquier, 1979). Other historians have been inclined to see the European marriage pattern arising as a response to pressures that built up on available resources as mortality declined in the phase of demographic recovery following the ravages of the plague-infested later Middle Ages (Chaunu, 1974: Livi-Bacci, 1978). The persistence of high I_m values in eastern and southern areas of Europe were therefore regarded as compatible with a diffusionist interpretation which perceives the 'new pattern' spreading outwards rather slowly from a north-west European cultural 'core region'. Such a view therefore presents the areas of low female marriage age and high marital intensity in the nineteenth century as 'survivals' of a medieval pattern that once prevailed over the whole of Europe.

While at odds with the chronology encountered in Goody's thesis on European marriage, the above position has also to confront evidence that is not wholly compatible with a diffusionist argument. Evidence from medieval England and from certain parts of southern Europe seems to suggest that fundamental features of marital patterns in these regions had been in place long before the sixteenth century and preceded the massive shifts in mortality or religious changes that characterized the two centuries after the arrival of bubonic plague in 1347. In England, late age and low incidence of female marriage seem to be attributes widely characteristic of society (outside the ranks of the landed aristocracy) from at least the thirteenth century, when it is also possible for the first time (because of data availability) to detect other correlates of the marriage pattern or household formation system, such as bi-lateral kinship, neolocal residence at marriage, a narrow age gap between spouses, a significant number of brides older than their husbands at first marriage, rather high levels of remarriage and, of particular importance, a high proportion (40–50%) of young adults of both sexes aged between 15 and 30 years living as servants in the households of non-kin (Smith 1979, 1981a, 1981b, 1983, 1986a; Goldberg 1986a, 1986b, 1988). Further evidence of the perdurance of marriage patterns, although in this case a very different pattern, is to be found in the detailed research into the remarkably rich archives of later medieval Tuscany in north-central Italy. This area, even before the Black Death and throughout the fifteenth century, possessed an early age of female first marriage (18 years), with almost 95% of women ever-marrying, a wide age gap between spouses (10–12 years), relatively low rates of female remarriage, a readily identifiable joint household formation system and a very limited place in the social structure for live-in 'life-cycle stage' servants or resident non-familial labour of any kind (Herlihy & Klapisch-Zuber, 1978; Smith, 1981a; Hajnal, 1982; Klapisch-Zuber, 1985: pp. 20–2; Smith, in press). There seem to be grounds for supposing that perduring contrasts of the kind that distinguished England from Tuscany were to be found elsewhere. For instance, evidence for south-east European areas from the fourteenth and fifteenth centuries points to the existence there of a marriage regime in many respects indistinguishable from that of Tuscany at comparable dates (Hammel, 1980; Laiou-Thomadakis, 1977; Rheubottom, 1988). Furthermore, the earliest nineteenth century Serbian censuses reveal a mean age of female marriage of 19 years and only a small percentage never-marrying (Hajnal, 1965: p. 109). These are features that

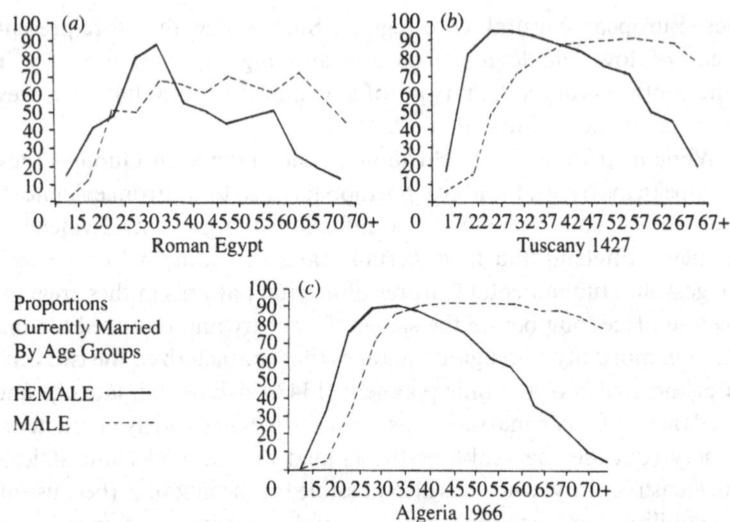

Figure 12.1. Nuptiality data from some Mediterranean populations.

distinguish those resident within a Serbian quarter of Belgrade in 1723 when 77% of females aged 15–19 were married and no women whatsoever aged 45–59 were described as never-married (Laslett & Clarke, 1972: p. 397).

There is reason to suppose, based on an admittedly small body of evidence, that the marital patterns of many societies situated close to the north shore of the Mediterranean shared little in common with much of northern and north-west Europe before the eighteenth century. What evidence is there to suggest that the African and Asian shores possessed nuptiality patterns that show them to have been wholly unlike the northern Christianized shore? Hesitantly, we can consider data from what might be regarded as an unacceptably small sample of census fragments relating to Roman Egyptians. These data concern 172 household census returns, containing 880 persons distributed between 8 towns and 20 villages for various points in time extending from A.D. 19 to A.D. 257. The age patterns of marriage are plotted in Figure 12.1(*a*). It should not be expected that with these rather small numbers that the line graphs will assume the smooth regularity encountered in far larger bodies of evidence. Nonetheless, the form displayed by these data is remarkably similar to that for a similar plot of evidence from the Florentine catasto of 1427 and the Algerian national census of 1960 (See Figures 12.1(*b*) and 12.1(*c*)). Hopkins (1980: p. 333), the author of a brilliant study of the Roman Egyptian evidence, primarily concerned with an assessment of brother–sister marriages, is inclined to regard these census fragments as pointing to 'a much later age for female marriage than we would expect, following Hajnal's path-breaking work, which identified a world pattern of early female marriage (before 20) in pre-industrial societies outside northwestern Europe and North America'. However, such a view of Roman Egyptian marriage fails to attach sufficient weight to the existence in these data of a number of very early marriages of females before age 14 and a marital incidence of 40% among females aged 15–19 years. Furthermore, by age 34, 100% of females had married. The absence of very youthful male marriage is noteworthy. Indeed the data strongly suggest a wide age gap between husband and wife and minimal levels of remarriage by females compared with males. These same attributes are readily detectable in the Algerian and Tuscan populations for very different moments in time (See Figures 12.1(*b*) and (*c*)). While other historical evidence bearing on nuptiality from north Africa is lacking there are data from the Mediterranean north shore that at least permit us to draw comparisons between Christian and non-Christian populations in that area. Evidence from south-east Europe recovered from Ottoman registers relating to both Christian and Muslim populations in Bulgaria suggests that female mean age at marriage was between 18 and 20 in the 1860s. These women, of both religious affiliations, took husbands who were approximately 10 years their senior (Todorova, in press).

In moving further back in time and to the western end of the Mediterranean basin, we should note that Vincent's (1974) investigations of the marriage patterns of the moriscos (Muslim converts to Christianity) in Estramadura in 1594 suggest no significant differences between Christianity and Islam; girls married early under both faiths. The mean age for morisco girls was 18.6 years. This figure is very close to that found in Pedralba in Valencia, among the Christian population in the second and third decades of the seventeenth century (Casey, 1979).

Another pioneering study by Hopkins (1964–65) resulted in the publication of some highly influential, much cited, estimates on the age of marriage of Roman females in the late Empire. The evidence is epigraphic and concerns inscriptions produced by men who commemorated their dead wives, giving both an age at death and a length of marriage. Hopkins noted that the findings from this evidence revealed that there was a tendency for the age at marriage to rise with the spread of Christianity. He concluded that the modal age at marriage of pagan girls was 12–15 years and that of Christian girls was 15–18. These data have recently been reworked and reassessed by Shaw (1987) who argues forcefully that the early age of pagan marriage is typical only of an elite group in Rome itself and is hardly representative of the broad mass of Roman society. In fact, outside of

Rome and outside this elite group, age at first marriage would seem to have been between 18 and 20 years for females. Shaw argues that the explanation for rather later age of Christian brides seems to be a function of their social status. Indeed, marriage patterns in the Christian sample, far from being peculiarly Christian is simply evidence of a broader lower-class form of nuptiality that typified most of Roman society during the earlier non-Christian centuries. If so, these patterns are likely to have been close to those displayed by the inhabitants of Roman Egypt, fifteenth-century Tuscany and Algeria in 1960. The evidence is fragmentary, and scattered both in space and time, but taken as a whole it throws substantial doubt on Goody's claim that the unity of the circum-Mediterranean area, in these fundamental respects, was broken as a result of processes set in train by Christianity's transition from a sect to a Church after the early fourth century; and in particular by its attempt to encourage exogamy, and to prohibit the practices which families had supposedly adopted so as to protect their properties from the kind of fragmentation which the existence of diverging devolution constantly threatened to bring about.

While a view of persisting patterns over many centuries can be entertained with some confidence for England and south-east Europe it is an argument that has to be developed cautiously. Indeed, recent research findings from some other parts of Europe, especially southern France, northern Iberia and north-central Italy, suggest that notable shifts in marital patterns may have taken place in the eighteenth and nineteenth centuries. In all of these latter regions case studies suggest a pattern of early marriage (18–22 years for female first marriages) and low proportions never marrying, in the years before and after 1350 and which is still identifiable into the early eighteenth century (Le Roy Ladurie, 1975; Rossiaud, 1976; Laribière, 1967; Higounet-Nadal, 1978; Landes-Mallet, 1985; Vincent, 1974; Anson Calvo, 1977; Bell, 1979; Corsini, 1971; Casey, 1979; McArdle, 1978; McLennan, 1980; Soler-Serratosa, 1985; Valero Lobo, 1984). However, it would seem that by the nineteenth century I_m, as revealed by the Princeton Fertility Project, suggest that there had been a significant rise in the age and a diminution in the incidence of female marriage (Perez Moreda, 1986; Reher & Iriso-Napal, 1989; Cachinero-Sanchez & Soler-Serratosa, in press; Rowland, 1988). Both the degree and the geographical extent of these changes remain the subject of current research, much as yet unpublished. However, provisional findings do raise the possibility of three broad early modern European zones. Two of these – the north-west and the extreme south of Europe – were characterized by only limited change in marital patterns that remained geographically differentiated over millenia[3]; whilst a 'transitional' zone underwent a shift upwards in the age of female marriage and a rise in definitive celibacy at various points between the fifteenth and early nineteenth centuries to constitute what Ansley Coale (1973) once called a 'malthusian' transition. There are grounds for supposing that large parts of Finland and the Baltic provinces of Latvia, Lithuania and Estonia located immediately to the west of Hajnal's fault-line in 1871 had 'changed sides' in the second half of the eighteenth century (Lutz & Pitkanen in press; Coale, Anderson & Harm, 1979).

To date, these developments in the 'transitional zone' have been interpreted largely in homeostatic terms. Nuptiality is conceptualized by those who advocate such an interpretation primarily as a process determined by underlying levels of mortality and adjusting so as to ensure that intrinsic growth rates did not greatly exceed zero per cent per annum. Indeed, such a conceptual framework has provided some scholars with an explanation of the European marriage pattern itself (Livi-Bacci, 1978). Those who argue in this fashion would wish to suggest that by the late eighteenth century I_m values over much of northern Spain, Portugal north of the Tagus, southern France, northern and central Italy rested somewhere below those encountered over the Portuguese Alentejo, Andalusia, Estramadura and Valencia in Spain and the Italian South (Sicily, Basilicata, Calabria, Apulia) and considerably below those found in the bulk of eastern and south-eastern Europe. The preferred view would be to explain the lowered level of nuptiality as being in a large measure the product of increased survivorship probabilities which thereby induced countervailing rises in marriage age and declines in marital incidence. An interpretation of this kind, while highly plausible and logically coherent, would seem incapable of accounting for those large areas of north-west Europe with I_m values well below 0.55 in 1870 and most likely for many centuries before, unless it could be argued that this region had been endowed with more moderate levels of mortality than other areas of the continent from at least the high middle ages. Furthermore, what we now know of patterns of marital change in England in the three and a half centuries after 1540 suggests the irrelevance of a model of nuptiality that gave mortality the pivotal role in determining the rate at which 'niches' were vacated and hence set the size of the openings through which marriages 'flowed into' the society (Wrigley & Schofield, 1981).

Demographers have been too ready to allocate the role of independent variable to mortality in their discussions of the relationship between life expectancy and nuptiality. As Wrigley (1987a) notes 'we are less accustomed to consider the possibility that the kind of social conventions well suited to coping with prevailing high levels of mortality may have themselves produced the situations to which they were well adapted when the group in question lived in an area in which mortality was not necessarily at a high overall level'. In this sense it may be reasonable to suppose that mortality was enhanced to counteract the fertility that was itself 'inappropriately' high given the population's broader ecological context. There would seem little reason to assume that it was only in north-west Europe that there was a particularly benign disease environment or agricultural systems so productive that the 'platform' level of mortality was low on account of favourable nutrition deriving from predictably adequate diets. Indeed, it is worth considering the likelihood of there having existed in that region a mortality regime which had been engineered by the social customs which themselves exercised such a strong influence on fertility levels and thus indirectly upon mortality. It is evident that in the eighteenth and nineteenth centuries mortality levels move within very different ranges in north-west and southern Europe; in the north-west life expectancy at birth (e_0) oscillates between 35 and 40, while in the south e_0 was frequently well below 30 or even 25 years. How far we are justified in seeing the low survivorship in southern Europe as related to this region's nuptiality and higher total fertility is a question that is easier to pose than to answer (Perez Moreda & Reher, 1985).

Attempts to understand European marriage characteristics and their associated fertility consequences through models that rely heavily on property, its mode of transmission, and its social distribution have had limited explanatory success. Those explanations of marriage patterns which treat mortality as an independent variable whose influence on nuptiality and hence fertility may be mediated through a particular system of resource allocation in a steady-state economy are possibly more successful, although they create geographical 'residuals' that render the model unconvincing as a general theory. Increasingly, in recent years it is possible to encounter scholars who prefer to treat the household-formation systems themselves, in particular the value systems affecting interpersonal relations within them, as the independent variable (Todd, 1985, 1987; Cain & McNicoll, 1988; Bruce and Dwyer, 1988).

It needs to be stressed that within the broad area of late and low intensity marriage for both sexes that extended over north-west and west Europe there were striking differences reflected in the way the 'nuptiality valve' or 'preventive check' operated. Recent work on Alpine demography in the eighteenth and nineteenth centuries is especially pertinent (Viazzo, 1989). A consensus seems to have emerged which stresses that the arrival of the potato in the Alpine regions (all largely within the area of Hajnal's European marriage pattern) led to a very significant improvement of the region's agricultural productivity after 1750. It is argued that the amelioration in diets that ensued resulted in falls in infant and child mortality and increases in the fecundity of marriages. The net result was an increase in population growth rates that led to a build-up of pressure on limited agricultural resources in this mountainous zone. It is indeed noteworthy that in 1870 the European Fertility Project findings reveal the Alps, broadly defined, to be in possession of very low fertility. In fact, I_f levels were frequently no more than 30% of those achieved by the Hutterites[4]. However, marital fertility was not in general distinguished by being unusually low. Quite the contrary; I_g frequently reached levels of 0.75 to 0.88. In many areas levels of I_m were below 0.40, apparently confirming Malthus' view that in the Alps 'the preventive check must have been unusually great'. But it was far from being the only manifestation of demographic homeostasis. Responses to the threats deriving from population growth initiated by mortality falls took on a variety of forms that while leading to low regional total fertility present us with a striking instance of equifinality.

In the Austrian Alps, population growth seems to have spawned a dramatic decline in nuptiality as I_m fell to levels of 0.30 or lower. Yet, overall fertility was no lower than elsewhere in the Alps since marital fertility was high on account of the practice of not breastfeeding at all or only for a short period, and illegitimacy was rising sharply. As Viazzo (1989: p. 192) notes the peasants in this region strictly adhered to a pattern of partible inheritance that revealed the social system to be 'brittle' in its homeostatic responsiveness. In fact, nuptiality appears to have been blocked well below the carrying capacity of the local economy in order to perpetuate a given kind of social structure (Lesthaeghe, 1980).

Yet in the Western Alps values of I_m were not noticeably depressed over large areas although total fertility was no higher than in the eastern regions. In many parts of Alpine France and Switzerland high levels of seasonal and temporary migration occurred and served to depress marital fertility as husbands were absent from the marital bed for long periods thereby intensifying the fertility-reducing effects of long-duration breastfeeding (Viazzo, 1989: pp. 192–202). Locally too, male emigration created very skewed sex-ratios that reduced the marital opportunities of women to such an extent that female celibacy could reach or exceed 50% (Van de Walle, 1975). Furthermore, in certain communities in the western Alps, particularly where complex households were encountered, young married couples began their married lives co-residing with the groom's father. In these situations authority was securely held by older persons who seem to have imposed fertility restraint on the junior generation who married at particularly young ages. As in the eastern regions the household system appears to have exercised an independent influence on fertility restraint (Viazzo, 1989: pp. 244–6).

While the Alpine evidence is generally consistent with the principle that there would be a strong response on the part of fertility to a significant rise in life expectancy, it has been intriguingly demonstrated that a range of reactions are detectable that reflected regional traditions that ran very deep in local social structures and certainly pre-dated the boost to resources and demographic growth provided by the arrival of the potato.

However, it is not the intention of this discussion to down-grade the role of the economy to that of an inconsequential determinant of marital behaviour. Indeed it was noted in the introduction to this paper that the economic pre-requisites of marriage in Western Europe were highly distinctive. Considerable resources were necessary to initiate a marriage: ability to rent a house, the cost of furniture, cooking and eating utensils and other equipment necessary to run it, the acquisition of working capital to equip a farm, were important elements that influenced the speed taken by couples on the road to marriage. Arthur Young (1771: pp. 110–25) writing of rural England, reckoned that in the late eighteenth century it would cost between £60 and £90 to begin life on a small farm of 12 to 16 acres. Without family support this required a long period of saving. We have already considered the possible role that life-cycle service in husbandry may have played in this savings process. A majority of each new cohort of young men and women in their teens and early twenties hired themselves out for annual contracts whereby they laboured in the fields and workshops and acquired skills they might well use later in life. While entry into service was commoner among children of the poor, it was not absent among the wealthy echelons of society (Kussmaul, 1981). Consequently, it was not necessarily economics that caused parents to put their children out to service, but acceptance of a social norm, a view of what constituted the normal transitional stage between dependent childhood at home and independent existence later in life. It should be stressed that this inter-household movement of children brought about a divergence between the private and the social rate of return in children (Smith, 1986b). It led to some parents paying the costs of rearing their children, but receiving the economic return from other people's children. Furthermore, the existence of the institution of life-cycle service could have 'dysfunctional' effects on demographic trends in certain contexts. If attention is placed upon the demand rather than the supply factors that determined the incidence of service, the reasons for this dysfunction can be readily appreciated. In England it would seem that demand for servant labour on the part of farmers increased when population growth was zero or negative and wages were relatively high due to labour shortages. Under such conditions farmers preferred to employ live-in servant labour paid disproportionately in food and lodging. However, in conditions when wages were low and populations were growing, farmers preferred to hire their labour by the day and remunerate with cash rather than in kind (Smith, 1981b). Servants in England tended to increase in their incidence when mortality was high and to exacerbate the demographic impact of reduced life expectancies by reducing nuptiality and hence fertility in the younger adult age groups (Smith, in press).

Roger Schofield (1989) has recently drawn what is surely a critical distinction between the social systems in north-west Europe on the one hand founded on familistic principles and those distinguished by deeply entrenched individualistic–collectivist principles. In the relatively undifferentiated economy of family farms and rural crafts and services economic activity primarily takes place within the household: the movement of capital (land and equipment) takes place largely through or within the kin group; parents are likely to be rather more heavily involved in marriage decision-making. Support for the elderly is largely provided by younger generations within the family. Under these conditions while the population may utilize the nuptiality valve to determine the relationship between fertility and resource usage, it tends to encourage a society with relatively limited geographical and social mobility.

On the other side of the spectrum there is to be found the more differentiated pre-industrial society in which a significant proportion of the population sells its labour to be applied to capital over which it has only limited control. There are still some whose access to land or capital is through inheritance, but in general, inter-generational property flows are of limited consequence. Most children are expected to leave home, accumulate savings, choose their own marriage partner, and set up privately funded households of their own. Population is mobile and the family is an inadequate or unwilling provider of welfare. Here is the apparent paradox: these basically individualistic systems necessarily co-exist with strong collectivist systems providing care of the elderly and the weak (Smith, 1984b, 1988).

The familistic system appears more readily to reveal a synchronized link between mortality and fertility, especially in rural societies endowed with fixed numbers of economic niches. Demographic behaviour in rural France from the seventeenth to nineteenth centuries seems to exhibit the characteristics of the homeostatic system *par excellence*, with fertility falling steadily to counteract the effect of steadily rising life expectancy (Wrigley, 1987b: chapter 11).

In the individualistic society, fertility is likely to be determined by influences that are mediated through markets, both domestic and international, and geographical movements that can be both internal and external; it is also susceptible to influences of welfare policy and policy shifts on the part of those who fund and manage welfare systems. As

Schofield (1989: p. 285) puts it, these systems are 'open to endogenous and exogenous influences'. We would not expect fertility in such a system necessarily to dance to a tune played by a nuptiality valve as it opened and shut in response to variations in mortality. Indeed 'exogenous' influences connected with inter-regional, inter-continental movements of peoples and diseases, variations in the sexual division of labour, incidence of life-cycle service and changing ideologies concerning the collectivist systems of welfare provision could individually or in unison create phases when fertility was held back even though mortality worsened or when fertility remained high or continued to increase as mortality risks diminished (Smith in press, a). A system containing such elements can only in a strained fashion be regarded as 'homeostatic' in that it was just as likely to exhibit positive as negative feed-back. It possessed the potential for both disastrously rapid population growth and surprisingly long-lasting demographic restraint or stagnation with concomitant phases of sharp declines in living standards and substantial rises in per capita incomes. Such lengthy periods of demographic 'stop and go' seem to be readily detectable in the record of the English demographic past from the thirteenth to nineteenth centuries.

Contrasts of this order which can be shown, certainly after 1550, to have existed between France and England, demonstrate that even north-west Europe possessed only limited demographic coherence in the past regarding the manner by which fertility regimes related to their economic 'space'. Of course, it is hard to resist the comment that these regions were all part of the area of Christian Europe which appears, on the basis of the evidence reviewed in this essay, from the first to the nineteenth centuries A.D., to have lacked demographic coherence of the kind predicted by models in which the mode of property inheritance is given the central, indeed pre-eminent, role as independent variable.

Notes

1 I_m is a measure of the contribution of marital status to the overall rate of childbearing; strictly speaking, it is a measure for a hypothetical population in which only married women are fertile, and in which married women are subject to maximum fertility rates at each age. I_m is the ratio of the number of births produced by married women in such a population to the number that would be produced if all women were married. It measures the extent to which marital status would limit childbearing if marital fertility were natural and non-marital fertility were zero.

2 I_g is the ratio of the number of births occuring to married women to the number that would occur if married women were subject to maximum fertility. Maximum fertility is defined as the births that would occur if women at each age experienced the rate of childbearing at that age in the most prolific population reliably recorded – the Hutterites 1921–30.

3 It is a matter of some interest that, although female marriage was generally early, there was certainly considerable variation in marital patterns in southern Italy. Delille (1985), in particular, has attempted to contrast relatively late marriage, low total fertility and moderate mortality in certain peasant communities found in the hills and mountains of this

area with early marriage, high birth rates and very low life expectancy in the plains where cereal-producing latifundia were economically dominant. In fact, he draws a provocative contrast between the 'high pressure' environments of 'corn demography' and the 'low pressure' systems of 'tree demography'. For a thoughtful overview of the historical evidence bearing on family and demography in southern Italy see Benigno (1989).

I_f is a measure of the fertility of all women in the population; it is the ratio of the actual number of births to the hypothetical number if women were subject to the Hutterite fertility schedule.

References

Anson Calvo, M. C. (1977). *Demografia y sociedad urbane en la Zaragoza del siglo XVIII.* Zaragoza: Biblioteca Jose Sinues.

Bell, R. M. (1979). *Fate and Honour, Family and Village: Demographic and Cultural Change in Rural Italy since 1800.* Chicago: University of Chicago Press.

Benigno, F. (1989). The Southern Italian family in the early modern period: a discussion of co-residence patterns. *Continuity and Change*, **4**, 165–94.

Berkner, L. & Mendels, F. (1978). Inheritance systems, family systems, and demographic patterns in Western Europe, 1700–1900. In *Historical Studies of Changing Fertility*, ed. C. Tilly, pp. 209–24. Princeton N.J.: Princeton University Press.

Bonfield, L. (1986). Normative rules and property transmission: Reflections on the link between marriage and inheritance in early modern England. In *The World we have Gained: Histories of Population and Social Structure*, ed. L. Bonfield, R. M. Smith & K. Wrightson, pp. 155–76. Oxford: Basil Blackwell.

Bruce, J. & Dwyer, D. (1988). Introduction. In *A Home Divided: Women and Income in the Third World*, ed. D. Dwyer and J. Bruce, pp. 1–19. Stanford: Stanford University Press.

Cachinero-Sanchez, B. & Soler-Serratosa, J. (1990). Nuptiality and celibacy in Spain at the end of the eighteenth century. In *Regional and Spatial Demographic Patterns in the Past*, ed. R. M. Smith, Oxford: Basil Blackwell (in press).

Cain, M. & McNicoll, G. (1988). Population growth and agrarian outcomes. In *Population Food and Rural Development*, ed. R. D. Lee, W. B. Arthur, A. C. Kelley, G. Rogers & T. N. Srinivasan, pp. 101–17. Oxford: Clarendon Press.

Caldwell, J. C., Caldwell, P. & Quiggin, P. (1989). The social context of AIDS in sub-Saharan Africa. *Population and Development Review*, **15**, 185–234.

Casey, J. (1979). *The Kingdom of Valencia in the Seventeenth Century.* Cambridge: Cambridge University Press.

Chaunu, P. (1974). *Histoire, science sociale. La durée, l'espace et l'homme à l'époque moderne.* Paris: Mouton.

Coale, A. J. (1973). The demographic transition. In *International Union for the Scientific Study of Population: International Conference, Liège*, vol. 1, pp. 53–72. Liège: International Union for the Scientific Study of Population.

Coale, A. J., Anderson, B. & Harm, E. (1979). *Human Fertility in Russia since the Nineteenth Century.* Princeton N.J.: Princeton University Press.

Coale, A. J. & Treadway, R. (1986). A summary of the distribution of overall fertility, marital fertility and the proportion married in the provinces of Europe. In *The Decline of Fertility in Europe*, ed. A. J. Coale and S. C. Watkins, pp. 31–181. Princeton N.J.: Princeton University Press.

Coale, A. J. & Watkins, S. C. (eds.) (1986). *The Decline of Fertility in Europe*. Princeton: Princeton University Press.

Corsini, C. (1971). Richerche di demografia storico vel territorio di Firenze. *Quaderni Storici*, (no vol. number), 371–98.

Czap, P. (1978). Marriage and the peasant joint family in era of serfdom. In *The Family in Imperial Russia*, ed. D. L. Ransel, pp. 103–23. Chicago: University of Illinois.

Delille, G. (1985). *Famille et propriété dans le royaume de Naples (XVe-XIXe siècles)*. Rome–Paris: Ecoles Francaise de Rome.

Derouet, B. (1980). Une démographie sociale différentielle. *Annales, Economies, Sociétés, Civilisations*, **35**, 3–41.

Drake, M. (1969). *Population and Society in Norway 1735–1865*. Cambridge: Cambridge University Press.

Drake, M. (1974). *Historical Demography: Problems and Projects*. Milton Keynes: Open University Press.

Dupâquier, J. (1979). Population. In *The new Cambridge modern history XII: Companion volume*, ed. P. Burke, pp. 80–114. Cambridge: Cambridge University Press.

Elliot, V. B. (1981). Single women in the London marriage market: Age, status and mobility, 1598–1619. In *Marriage and society: Studies in the social history of marriage*, ed. R. B. Outhwaite, pp. 81–100. London: Europa Publications Limited.

Fitzpatrick, D. (1985). Marriage in post-famine Ireland. In *Marriage in Ireland*, ed. A. Cosgrove, pp. 116–31. Dublin: College Press.

Goldberg, P. J. P. (1986a). Marriage, migration, servanthood and life-cycle in Yorkshire towns of the Middle Ages: Some York cause paper evidence. *Continuity and Change*, **1**, 141–69.

Goldberg, P. J. P. (1986b). Female labour service and marriage in the late medieval urban North. *Northern History*, **22**, 18–38.

Goldberg, P. J. P. (1988). Women in fifteenth-century town life. In *Towns and Townspeople in the Fifteenth Century*, ed. A. F. Thompson, pp. 101–28. Gloucester: Alan Sutton.

Goody, J. R. (1969). Inheritance, property and marriage in Africa and Eurasia. *Sociology*, **3**, 55–76.

Goody, J. R. (1971). Class and marriage in Africa and Eurasia. *American Journal of Sociology*, **76**, 585–603.

Goody, J. R. (1973a). Bridewealth and dowry in Africa and Eurasia. In *Bridewealth and Dowry*, ed. J. R. Goody & S. J. Tambaih, pp. 1–58. Cambridge: Cambridge University Press.

Goody, J. R. (1973b). Polygyny, economy and the role of women. In *The Character of Kinship*, ed. J. R. Goody, pp. 175–89. Cambridge; Cambridge University Press.

Goody, J. R. (1976). *Production and Reproduction: A Comparative Study of the Domestic Domain*. Cambridge: Cambridge University Press.

Goody, J. R. (1983). *The Development of the Family and Marriage in Europe*, Cambridge: Cambridge University Press.

Goody, J. R. & Buckley, J. (1973). Inheritance and woman's labour in Africa. *Africa*, **43**, 108–21.

Goody, J. R., Irving, B. & Tahany, N. (1971). Causal inferences concerning inheritance and property. *Human Relations*, **24**, 295–314.

Goody, J. R. & Tambiah, S. J. (1973). *Bridewealth and dowry*. Cambridge: Cambridge University Press.

Hajnal, J. (1965). European marriage patterns in perspective. In *Population in History: Essays in Historical Demography*, ed. D. V. Glass & D. E. C. Eversley, pp. 101–47. London: Edward Arnold.

Hajnal, J. (1982). Two kinds of pre-industrial household formation system. *Population and Development Review*, **8**, 449–84.

Hammel, E. A. (1980). Household structure in fourteenth-century Macedonia. *Journal of Family History*, **5**, 242–73.

Herlihy, D. & Klapisch-Zuber, C. (1978). *Les Toscans et leurs familles*. Paris: Presses de la Fondation Nationale des Sciences Politiques.

Higounet-Nadal, A. (1978). *Perigueux aux XIVe et XVe siècles*. Bordeaux: Fédération historique du Sud-Est.

Hoch, S. L. (1986). *Serfdom and Social Control in Russia: Petrovskoe, a village in Tambov*. Chicago: University of Chicago Press.

Hopkins, K. (1964–65). The age of Roman girls at marriage. *Population Studies*, **18**, 309–27.

Hopkins, K. (1980). Brother–sister marriage in Roman Egypt. *Comparative Studies in Society and History*, **22**, 303–55.

Klapisch-Zuber, C. (1985). *Women, Family and Ritual in Renaissance Italy*. Chicago: University of Chicago Press.

Knodel, J. (1988). *Demographic Behaviour in the Past: A Study of Fourteen German Village Populations in the Eighteenth and Nineteenth Centuries*. Cambridge: Cambridge University Press.

Kussmaul, A. (1981). *Servants in Husbandry in Early Modern England*. Cambridge: Cambridge University Press.

Laiou-Thomadakis, A. E. (1977). *Peasant Society in Late Byzantine Empire: A Social and Demographic Study*. Princeton: Princeton University Press.

Landes-Mallet, A. M. (1985). *La famille en Rouergue au moyen age, étude de la pratique notariale*. Rouen: Publication de l'Université de Rouen.

Laribière, G. (1967). Le mariage à Toulouse aux XIVe et XVe siècles. *Annales du Midi*, **79**, 335–61.

Laslett, P. & Clarke, M. (1972). Houseful and household in an eighteenth-century Balkan city. A tabular analysis of the listing of the Serbian sector of Belgrade in 1733–4. In *Household and Family in Past Time*, ed. P. Laslett, pp. 375–400. Cambridge: Cambridge University Press.

Le Roy Ladurie, E. (1975). *Montaillou, village occitan de 1294 à 1324*. Paris: Gallimard.

Lesthaeghe, R. (1980). 'On the social control of human reproduction'. *Population and Development Review*, **6**, 527–48.

Livi-Bacci, M. (1978). *La société italienne devant les crises de mortalité*. Florence: Dipartimento Statistico.

Lutz, W. & Pitkanen, K. (in press). Tracing back the eighteenth-century 'Nuptiality transition' in Finland. In *Regional and Spatial Demographic Patterns in the past*, ed. R. M. Smith. Oxford: Basil Blackwell.

McArdle, F. (1978). *Altopascio: A study in Tuscan rural society 1587–1784*. Cambridge: Cambridge University Press.

McLennan, J. (1980). A demographic study of St. Paul's parish Valetta, Malta 1598–1798, using the method of family reconstitution. Unpublished University of Aberdeen Ph.D. Thesis.

Monter, E. W. (1979). Historical demography and religious history in sixteenth-century Geneva. *Journal of Inter-disciplinary History*, **9**, 399–427.

Perez Moreda, V. (1986). Matrimonio y familia algunas considerraciones sobre el modelo matrimonial espanol en la edad moderna. *Boletin de la Asociacion de Demografia Historica*, **4**, 3–51.

Perez Moreda, V. & Reher, D. S. (1985). Demographic mechanisms and long-term swings in population in Europe. In *Proceedings of the International Population Conference of the International Union for the Scientific Study of Population Florence 1985*, vol. 4, pp. 313–29. Liège: International Union for the Scientific Study of Population.

Reher, D. S. & Iriso-Napal, P. L. (1989). Marital fertility and its determinants in rural and urban Spain, 1887–1930. *Population Studies*, **43**, 405–27.

Rheubottom, D. B. (1988). 'Sisters first': Betrothal order and age at marriage in fifteenth-century Ragusa, *Journal of Family History*, **13**, 359–76.

Rossiaud, J. (1976). Prostitution, jeunesse et société dans les villes du Sud-est au XVe siècle. *Annales Economies, Sociétés, Civilisations*, **31**, 289–325.

Rowland, R. (1988). Sistemas matrimoniales en la Peninsula Iberica (siglos xvi–xix). Una perspectiva regional. In *Demografia historica en espana*, ed. V. Perez-Moreda & D. S. Reher, pp. 74–137. Madrid: Ediciones el arquero.

Russell, J. C. (1948). *British Medieval Population*. Albuquerque: The University of New Mexico Press.

Schofield, R. S. (1985). English marriage patterns revisited. *Journal of Family History*, **10**, 2–20.

Schofield, R. S. (1989). Family structure, demographic behaviour and economic growth. In *Famine, Disease and the Social Order in Early Modern Society*, ed. J. Walter & R. S. Schofield, pp. 279–304. Cambridge: Cambridge University Press.

Sharlin, A. (1977). Historical demography as history and demography, *American Behavioural Scientist*, **21**, 245–62.

Shaw, B. D. (1987). The age of Roman girls at marriage: some reconsiderations. *Journal of Roman Studies*, **77**, 30–44.

Smith, R. M. (1979). Some reflections on the evidence for the origins of the 'European marriage pattern' in England. In *The Sociology of the Family*, Sociological Review Monograph 28, ed. C. Harris, pp. 74–112. Keele: University of Keele.

Smith, R. M. (1981a). The people of Tuscany and their families: medieval or mediterranean?. *Journal of Family History*, **6**, 107–28.

Smith, R. M. (1981b). Fertility, economy and household formation in England over three centuries, *Population and Development Review*, **7**, 595–622.

Smith, R. M. (1983). Hypothèses sur la nuptialité en Angleterre aux XIIIe–XIVe siécles. *Annales, Economies, Sociétés, Civilisations*, **38**, 107–36.

Smith, R. M. (1984a). Some issues concerning families and their property in rural England. In *Land, Kinship and Life-cycle*, ed. R. M. Smith, pp. 1–86. Cambridge: Cambridge University Press.

Smith, R. M. (1984b). The structured dependence of the elderly as a recent development: Some sceptical historical thoughts. *Ageing and Society*, **4**, 409–28.

Smith, R. M. (1986a). Marriage processes in the English past: Some continuities. In *The World we have Gained: Histories of Population and Social Structure*, ed. L. Bonfield, R. M. Smith & K. Wrightson, pp. 43–99. Oxford: Basil Blackwell.

Smith, R. M. (1986b). Transfer incomes, risk and security: The roles of the family and the collectivity in recent theories of fertility change. In *The State of Population Theory*, ed. D. A. Coleman and R. Schofield, pp. 188–211, Oxford: Basil Blackwell.

Smith, R. M. (1988). Welfare and the management of demographic uncertainty. In *The Political Economy of Health and Welfare*, ed. M. Keynes, D. A. Coleman & N. H. Dimsdale, pp. 108–35. London: Macmillan.

Smith, R. M. (in press, a). Exogenous and endogenous influences on the 'preventive check' in England 1600–1750: Some specification problems. *Economic History Review*.

Smith, R. M. (in press, b). Work and reputation: The economic activities of unmarried females in the household formation systems of northern and southern Europe in the later Middle Ages. In *The Historical Roots of the Western family; The Evolution of Family Relations in Italy*, ed. D. Kertzer & R. Saller. New Haven, Conn.: Yale University Press.

Soler-Serratosa, J. (1985). Demografia y sociedad en Castilla la Nueva durante el Antiquo Réegimen: la villa de Los Molinos, 1620–1730. *Revista Espanola de Investigaciones Sociologicas*, **32**, 141–92.

Sundt, E. (1980). *On Marriage in Norway*. trans. and introd. M. Drake. Cambridge: Cambridge University Press.

Todorova, M. N. (in press). Marriage and nuptiality in Bulgaria during the nineteenth century.

Todd, E. (1985). *The Explanation of Ideology: Family Structures and Social Systems*. Oxford: Basil Blackwell.

Todd, E. (1987). *The Causes of Progress: Culture, Authority and Change*. Oxford: Basil Blackwell.

Valero Lobo, A. (1984). Edad media de acceso al matrimonio en Espana. Siglos XVI–XIX. *Boletin de la Asociacion de Demografia Historica*, **2**, 39–48.

Van de Walle, F. (1975). Migration and fertility in Ticino, *Population Studies*, **29**, 447–62.

Viazzo, P. P. (1989). *Upland Communities, Environment, Population and Social Structure in the Alps since the Sixteenth Century*. Cambridge: Cambridge University Press.

Vincent, B. (1974). Les morisques d'Estrémadura au XVIe siècle. *Annales de Démographie Historique*, 431–38.

Watkins, S. C. (1986). Regional patterns of nuptiality in Western Europe, 1870–1960. In *The Decline of Fertility in Europe*, ed. A. J. Coale & S. C. Watkins, pp. 314–36, Princeton N.J.: Princeton University Press.

Weir, D. R. (1984). Rather never than late: Celibacy and age at marriage in English cohort fertility, *Journal of Family History*, **9**, 340–54.

Wrigley, E. A. (1978). Fertility strategy for the individual and the group. In *Historical studies of changing fertility*, ed. C. Tilly, pp. 135–54, Princeton N.J.: Princeton University Press.

Wrigley, E. A. (1985). *The local and the general in population history*. Exeter: University of Exeter.

Wrigley, E. A. (1987a). No death without birth: The implications of English mortality in the early modern period. In *Problems and Methods in the History of Medicine*, ed. R. Porter & A. Wear, pp. 133–50, London: Croom Helm.

Wrigley, E. A. (1987b). *People, Cities and Wealth: The Transformation of Traditional Society*. Oxford: Basil Blackwell.

Wrigley, E.A. & Schofield, R.S. (1981). *The Population History of England 1541–1871: A Reconstruction*. London: Edward Arnold.
Young, A. (1771). *The Farmer's Guide in Hiring and Stocking Farms*. 2 vols. London.

Index

page numbers in italics refer to figures and tables